ENERGY, ELECTRIC POWER, AND MAN

TIMOTHY J. HEALY

University of Santa Clara

BOYD & FRASER PUBLISHING COMPANY

3627 Sacramento Street, San Francisco, California 94118

Timothy J. Healy: **ENERGY, ELECTRIC POWER, AND MAN**

Library of Congress Catalog Card Number: 73–86050

ISBN: 0–87835–041–1

2 3 · 6 5 4

ENERGY, ELECTRIC POWER, AND MAN

DEDICATION

This book is dedicated to my wife, to my children, and to my children's children to come in the hope that they may see in their lifetimes solutions to the dilemma posed here. We in our generation and our children in theirs must find a compromise between our need for energy and the problems which come in meeting those needs. This book is written with confidence that those solutions will be found.

● ● ●

It is a futile task to try to single out all those who have contributed to this book. No person helped more to give the book style and continuity than Mr. Arthur Weisbach, who edited the manuscript. My thanks go to Mr. Ian Murray, Mr. Richard Pefley, Dr. Richard Dorf, and a number of other reviewers who read parts of the manuscript, and suggested many important changes. I received invalu-

able aid from the Sierra Club, and from many parts of the power
industry, in particular the Pacific Gas and Electric Company and
the Southern California Edison Company. Many more persons and
groups contributed very significantly. And finally my deep appreci-
ation to Dean Robert J. Parden of the School of Engineering at
the University of Santa Clara, without whose original suggestion
and continuing encouragement and support this book could not
have been written.

PREFACE

At approximately 5:16 P.M. on November 9, 1965, in a power station in Ontario, Canada, a protective relay opened to disconnect a transmission line carrying electric power to Toronto. This triggered a complex sequence of overloads and line disruptions which darkened neighboring regions within seconds. Within about twelve minutes 80,000 square miles and thirty million people in the United States and Canada were blacked out. The outage lasted from a few moments to over thirteen hours. Elevators and subways in New York stopped; air–traffic control systems failed; some hospitals lost power. Suddenly people learned what it meant to live without electric power. Some ate by candlelight, read forgotten books, and went to bed cold. The occasion inspired a number of jokes; a persistent but probably incorrect rumor that nine months later there was an abrupt rise in the birth rate, and a movie, "Where Were You When the Lights Went Out?" starring Doris Day and Robert Morse.

Five years later and 3,000 miles away, on a Sunday in the summer of 1970, 2,500 people met in a remote valley of southern

British Columbia to protest the proposed flooding of the valley by a hydroelectric project. They picnicked; they listened to speeches against the project, and they heard a protest song written about that valley—Skagit Valley—by the noted folksinger and composer Malvina Reynolds.

How is a protest in a remote canyon in British Columbia related to a blackout in New York? For the student of today's problems, they dramatize the dilemma which is the subject of this book: man's need for energy, as opposed to the problems associated with generating and using that energy. Man has succeeded, perhaps too well, in "taming nature," and he has come to realize that he must ask not only what he wants but what it really costs as well.

There are many ways to approach the problem of energy demand and supply. We have chosen to emphasize the problems associated with generating electric energy. Two factors motivate this decision. First, electric energy demand is increasing at twice the rate (seven percent a year) of total energy (three to four percent per year). Hence it is probably our most significant energy growth problem. Second, the difficulties relating to electric energy growth either involve total energy growth problems, or are representative of such problems. We believe that to understand the problems of electric energy demand and supply is to understand what has been called the "energy crisis."

The reader should keep in mind, as he goes through this book, that while the electric energy dilemma is our focus of study, the dilemma of total energy supply and demand must eventually command our attention. Accordingly, the climax of the book is a pair of chapters which consider total energy sources and use. Chapter 13 is a major in-depth study of all of our major sources of energy. Chapter 14 focuses on the greatest single user of energy—the automobile—and considers the potential for powering the automobile with electric energy.

The subject of the book is *Energy*; the focus for study is *Electric Power*, and the ultimate interest which motivates our study is the good of *Man*.

Two hundred years into the Industrial Revolution we are beginning to ask seriously whether our use of energy has always been wise. Does the good we have accomplished justify the undesirable and often unanticipated side effects? In many cases the answer must certainly be yes. In some cases it is no. But posing the question in today's society requires that we consider how man uses energy and what side effects arise from its use. The question of environmental impact will therefore be an important part of our study.

Before man can use electric energy he must generate it. Almost coincident with the realization that we might question our use of energy came the realization that the generation of electric energy can cause serious problems. There is a growing shortage of certain fuels. There are important increases in costs from inflation, rising labor costs, and the costs of sophisticated environmental–protection devices. There are delays in power plant construction caused by concern about environmental impact, new licensing procedures, and complex construction problems. These factors and others have led to what has come to be called an "energy crisis."

In the chapters ahead we shall outline the dimensions of the crisis. In the face of an energy crisis, we seek a compromise. Some say that there can be no compromise with pollution. On the contrary, to exist is to pollute. Man has always been a source of pollution and he always will be. We can only ask what level of pollution we will accept for the good to be achieved. It is not an easy question to answer! There is, however, a consensus in the United States that we must learn to control our resources and our use of energy more wisely than we have in the past. And there is a growing sense that we must act rapidly to put such control into effect.

Intelligent compromise requires knowledge. The purpose of this book is to help the reader learn about this challenge, which is at the same time technical and social, economic and political. The book should permit the reader to understand the various dilemmas and perhaps to make intelligent decisions about their resolution. Very little mathematical sophistication is necessary to understand the text. The reader can skip the few equations without substantially decreasing his understanding. He should, however, learn the basic terminology; the units of energy and power; the sizes of power plants, and the amount of energy required to do certain jobs. At first this may seem to be unnecessary memorization. Later, however, it should become clear that to know, appreciate, and use this material effectively requires a certain acquaintance with terms and magnitudes.

To accommodate a wide variety of reader interests and abilities, we have developed three types of problems at the end of each chapter, wherever they are appropriate. The first are *General Problems*. These require some fairly simple arithmetic and at most some elementary algebra. The second kind of problems are *Advanced Mathematical Problems*. They are designed to challenge the student with a background in advanced mathematics, including calculus. Finally, we have included some *Advanced Study Problems* for the more serious student. They sometimes will require some mathematical sophistication, but more commonly they simply require critical study beyond the contents of the book.

The first drafts of this text have been tested in classrooms for two years. Students have ranged from freshman humanities majors to senior engineers. The author (and teacher) has not observed a significant difference in the interest or ability to learn the material between such diverse groups.

We do not presume nor intend to suggest a single solution to the dilemmas presented here, although our sympathies may show through on occasion. We would hope, rather, to present the problems as they stand today and as completely and as fairly as possible. The solutions to problems of this magnitude do not and cannot come from any individual. Rather, they evolve over the course of time from the knowledge and the wisdom of society. The quality of that evolution depends in no small way on the quality of that knowledge and wisdom to which this book is dedicated.

CONTENTS

1

MAN'S USE OF ELECTRIC ENERGY

How use doth breed a habit in a man.

Shakespeare
The Two Gentlemen of Verona

1.1 A CHALLENGE FOR OUR TIME

In today's highly developed technological society man uses energy in almost unlimited ways for a wide variety of purposes including transportation, heating, recreation, industry, and commerce. Today about 25 percent of that energy goes into the production of electric energy. This proportion may increase to 50 percent by the end of this century.

The purpose of this book is to introduce the problems related to generating the electric energy which will be required in the years ahead. Unfortunately, all of the known ways of generating electric energy produce some undesired side effects. These include expending limited fuel reserves; heating of the environment (particularly our waters); preemption of desirable lands such as mountain valleys, ocean shores, and lake fronts; air and water pollution; and radioactivity.

We face a challenging dilemma; on one hand we have a huge projected increase in demand of perhaps five to 10 times our present requirements by the year 2000. Every known solution is undesirable in some ways. Resolving the dilemma of demand and supply will constitute one of the great social–technological challenges of our time.

We describe all the important generation technologies in the chapters that follow. We also look at their side effects. The time when engineers, planners, or citizens could afford to ignore the effects of their projects on the environment is past. We also investigate costs. The comparison of benefit and cost is our basic means for making decisions relating to the flow of goods between people. We also discuss politics, because we make decisions for large groups of people through political institutions. We also investigate use. Only when people understand where and in what quantities electric energy is used can they make intelligent decisions about its use in the future.

1.2 WORK, ENERGY, AND POWER

Three important terms which we use throughout this book are *work, energy,* and *power.*

Work can be defined as the product of the *force* acting on a body and the *distance* which the body moves in the direction of this force. We shall rely here largely on the reader's intuitive sense of physical work. Work is done when an object is lifted from the ground, or when an automobile is driven along a road, or when a lamp lights a room, etc. We shall attempt to develop that intuitive sense of work through some examples later in this section.

Energy is defined as the available capacity of a body to do work because of its position or its condition. The extent to which energy is available depends on the nature of the process. Work and energy have the same units of measure.

Power is defined as the time rate at which energy is made available or work is done. The faster you make energy available or do work the greater is the power.

Suppose you push a block up a frictionless plane, as in Figure 1-1. The energy stored when the block is at the top equals the work you did to get it there. This stored energy depends on the weight W and the height H, and not on how fast you push. But the power you generate in order to move the block does depend on how fast you push it. If you push slowly, and hence take a long time to do the work, the power is low. But if you push the block up rapidly the power is greater.

As another example, consider a lump of coal. The lump has a certain amount of energy—a certain capacity to do work—which depends on its nature or condition, and on the process used to make the energy available. If we specify these two factors, we can give the amount of energy which can be made available. But we cannot talk sensibly about the power contained in a lump of coal. Power depends on how fast the lump burns and hence how fast the energy is made available.

The most commonly-used measuring unit of electric power is the *watt*. To get an idea of how much power a watt measures, consider the power required to operate some typical household devices. See Table 1-1.

TABLE 1-1.

Power Requirements of Household Appliances	
Device	Power (Watts)
Light bulb (average size)	100
Color television	350
Toaster	1,000
Range (all elements on)	12,000

The watt is a fairly small unit. In electric power studies we often deal with the *kilowatt*, which is equal to 1,000 watts. Kilowatt is abbreviated KW.

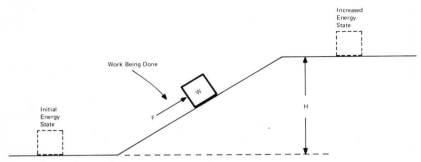

FIGURE 1–1. Work, Energy, and Power.

1,000 watts = 1 kilowatt = 1 KW

The kilowatt or KW rating of a device tells us at what time rate the device is using or will use energy, but it does not say how much energy will be required to do any particular job. We find the total energy required by multiplying power (the time rate of using energy) by time. We might use any time unit, but the hour is the usual choice. Therefore, the common unit of electric energy is one kilowatt–hour, abbreviated KWH.

Example 1-1

What is the cost of running a 100–watt light bulb continuously all month if the cost of electricity is two cents per KWH, which is close to an average residential rate? (There are 30 × 24 = 720 hours in a month.)

Energy = 720 × 100 = 72,000 watt-hours = 72 KWH
Cost = 72 × $0.02 = $1.44

Table 1-2 shows the amount of electric energy required for some specified activities.

Example 1-2

If energy costs $0.024 per KWH, how much does it cost to make a piece of toast, or to cook a large roast?

Toast: $0.024 × 1/30 = $0.0008
Roast: $0.024 × 5 = $0.12

TABLE 1-2.

Energy Required for Some Household Activities

Activity	Energy (KWH)
Run a 100 W light bulb for 10 hours	1
Watch color TV for two hours	7/10
Make a piece of toast	1/30
Cook a large roast	5

1.3 PRESENT POWER AND ENERGY REQUIREMENTS

How much power do we demand, and how much energy do we use in this country? (Our studies will be restricted largely to the United States for convenience. Figures for other countries will be smaller, but use patterns and growth patterns will be similar in many respects.)

Let's consider power first. The total stationary electric power generating capacity of the United States on January 1, 1972 was approximately 361,000,000 KW, or 361,000 MW (MW = 1,000 KW).

If we divide by our population (208,000,000) we see that we have available about 1.75 KW for each person.

The amount of total power demand depends on the number of tasks we wish to perform simultaneously. If all the devices which require electric power were turned on at once, the demand would greatly exceed the available supply of power, and protective equipment would shut off the power to protect the system. Fortunately, people spread their use of power devices at different times. While some are using devices such as ovens, refrigerators, washers, etc., others may be reading or listening to radios which demand little power. There are, however, certain periods when many people wish to perform tasks at the same time. An electric system must have sufficient power capacity to meet such peak demands. We shall discuss this question later.

In 1971 the energy used in the United States was approximately 1,600,000,000,000 KWH = 1.6×10^{12} KWH = 1.6 trillion KWH.

Again dividing by the population, we find that the average amount of energy used by each person in the United States in

1971 was about 7,700 KWH. This energy was not all used in homes, of course. Some was used in the manufacture of goods, some in selling goods, and some in other businesses. We shall return later to the distribution of electrical energy for its many uses.

Example 1-3

What percentage of the theoretically-available electric energy was actually used in 1971?

Find the approximate theoretically-available energy by multiplying the January 1, 1972 power capacity by the number of hours in 1971. ($365 \times 24 = 8,760$).

$$360,000,000 \times 8,760 = 3,150,000,000,000 \text{ KWH}$$

The percentage actually used was

$$\frac{1,600,000,000,000}{3,150,000,000,000} \times 100 = 51\%$$

At first glance it may seem that we have about twice as much capacity as we need. But that would be true only if our demand for power were constant. Since our demands are not constant, varying as they do with the number of tasks people wish accomplished, we need the greater power capacity.

1.4 FUTURE ENERGY REQUIREMENTS

In the previous section we considered our present requirements. In this section we consider where we have been and where we seem to be going with respect to electric power generation. In a sense this section is a sort of focal point for the entire book, because a huge projected increase in demand provides us with our present dilemma and our challenge for the future. We consider only energy in this section. Power requirements follow a similar growth curve.

The growth in electric energy use from 1948 to 1970, and the projected growth to the year 2000, are shown in Figure 1-2.

The projected forecast curves (dashed in Figure 1-2) assume that energy use increases at a constant rate of either seven percent

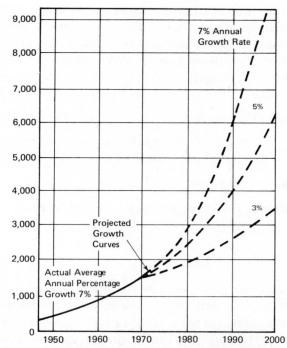

FIGURE 1—2. Electric Energy-Use Growth Curve (1948–2000).

per year, five percent per year, or three percent per year. (A curve which has a constant *rate* of growth is called an *exponential growth curve*. This is one of the most common and hence one of the most important types of curve; it appears in such widely separated areas as population growth, growth of capital invested with interest and compounded periodically, and discharge of an electric capacitor.)

Exponential growth occurs in any situation in which growth is proportional to the present size of the growing entity. This is, for example, a common growth mode for many human and animal populations as long as conditions for growth, such as food and space, remain adequate. However, when external factors such as famine, disease, or crowding begin to become significant, growth tends to level off. The graph will produce an S curve.

Growth in the electric power industry has been essentially exponential since the birth of the industry. How much will the industry grow if the present growth rate of about seven percent remains constant? Suppose we call the present electric energy level E_0 and assume that energy growth each year is r times the amount of energy for the given year. If we start with energy E_0, at the end of a year the energy is

$$E_1 = E_0 + rE_0 = E_0(1 + r)$$

So we start the second year with $E_0(1 + r)$. The rate is still r, so the energy at the end of the second year is

$$E_2 = E_1 + rE_1 = E_1(1 + r) = E_0(1 + r)^2$$

We can easily show that the energy at the end of n years is

$$E_n = E_0(1 + r)^n \qquad (1\text{-}1)$$

Example 1-4

Assume that the power industry continues to grow at a rate of 7% (r = 0.07). What will be its growth factor at the end of (a) 10 years; (b) 20 years; (c) 30 years; (d) 100 years? We need to find $(1 + 0.07)^n = (1.07)^n$. This can be done with a slide rule or with log tables.

a. $(1.07)^{10} = 1.97 \cong 2$
b. $(1.07)^{20} \cong 4$
c. $(1.07)^{30} \cong 8$
d. $(1.07)^{100} \cong 2^{10} = 1{,}024$

The example above illustrates that the growth *factor* is the same for any interval of k years, regardless of where the k years occur. From the example above you can see that a seven percent growth rate leads to a doubling period of ten years. Hence in 20 years, the factor is 2 × 2 = 4; in 30 years 8; 40 years 16, etc. If growth rate is constant, the number of years required for doubling is approximately 70/r. Hence, for example, if the growth rate is five percent, the doubling time is about 14 years.

If the growth rate continues, we shall be using about eight times as much electric energy in the year 2000 as in 1970. This is an

extremely large growth for this 30-year period. A number of predictions have been made about growth in the next 30 years.(1)
Most of these forecast a little less rapid growth. Typical forecasts
are for a growth of four to seven times. Any of these figures
four, or five, or six, or seven, or eight times, will require a tre-
mendous growth in the electric power industry by the year 2000.
It is this very dramatic growth projection, together with a realization of the difficulties associated with meeting this potential demand, which justify the generally accepted belief that we are in
the midst of or very near an "energy crisis."

Might leveling-off factors come into play in the years ahead?
We don't know, of course, what will happen, but we do know
that the present concern for the environment has resulted in some
questions about the need for a seven percent annual growth rate.
Some have suggested that cost rates be changed to discourage,
rather than encourage, use of large amounts of energy. Others have
suggested restrictive legislation to limit certain uses of electric
power. Such legislation, if directed only at the electric power
industry, could simply cause a shift of energy use from electric
energy to some other form. This might not be generally desirable.
There has been much discussion of, and agitation for, establishing
a comprehensive, explicit and consistent energy policy in the
United States. There is now an implicit policy which operates
through numerous government regulations and controls. It is probably time for a new general energy policy.

While it is not clear what may actually influence growth in the
future, it does seem clear that we shall continue to have some substantial measure of growth for the foreseeable future. We discuss
in this book all the apparent possible alternative schemes for meeting this growth demand. The book also examines the impact which
each alternative provides, both economically and environmentally.
It is our premise that a clear understanding of alternatives is necessary if we are to make intelligent decisions about how we should
increase electrical power generation.

But before we leave this growth rate question, we must ask how
much of the problem is attributable to population growth. The rate
of population growth is about one percent per year. Clearly, there

must be a yearly increase in energy *consumption* per person per year. We can express a net growth factor as the product of growth factors from separable effects. Calling r_p the population growth rate, and r_c the per capita energy–use growth rate, we have

$$(1 + r) = (1 + r_p) (1 + r_c) \qquad (1\text{-}2)$$

Solving (1-2) for r_c yields

$$r_c = \frac{r - r_p}{1 + r_p} \qquad (1\text{-}3)$$

For r_p = one percent and r = seven percent we obtain an r_c of about six percent. Hence the per capita use growth rate is almost six times that of the population. This shows us immediately that reducing the population growth rate to zero would not greatly affect the electric power growth problem.

1.5 HOW DO WE USE ELECTRIC ENERGY?

It is impossible to answer this question completely. The use of electric energy is far too broad and extensive. We can, however, indicate how much is used by each of the major sectors of society, and we can give a number of examples of use in particular situations. From this information we can see a pattern which will give us a good impression of how we use our electric energy.

It is common to make a four–fold grouping of electric–energy users. Table 1-3 gives the use for 1969.(2)

TABLE 1–3.

Electric Energy Users

Classification	KWH (million)	Percent
Industrial	557,220	43
Residential	407,922	31
Commercial	286,686	22
Other	55,350	4
Total	1,307,178	100

The industrial group includes manufacturing plants, textile, steel, lumber, mining, paper, and many other industrial processes. Residential includes electric energy used in homes and apartments for living purposes. Commercial refers to the energy necessary for running offices, stores, and shops. "Other" includes railroads, street lighting, government facilities and many more categories of users.

The percentage figures above are fairly consistent from one area to another, although a particular region or service area may emphasize one of them somewhat, depending on the local mix of homes and businesses.

Our next task is to look behind these gross numbers to find some specific examples of energy use within the groups. First take residential use, since it is familiar to most of us.

Table 1-4 gives the power required and the average energy used by some residential energy consumers.(3) (It is an extension of Tables 1-1 and 1-2.) The reader will no doubt find it interesting to spend some time looking at these numbers, and perhaps looking for some characteristic patterns or significant points hidden among the raw statistics. You may wish to interrupt your reading of the text at this point to look for such points for yourself before we make some observations of our own below.

The average *household* in the United States uses about 7,300 KWH each year. (This is consistent with our earlier figure of 7,700 KWH per *person*, combined with a household occupancy of a little over three persons and the 30 percent residential use figure above.) It is apparent that any household with all of the devices in Table 1-4 is using much more energy than one might expect from the average. In fact, the figure for electric heating is 16,000 KWH by itself. Most homes do not use electric heating; perhaps this is fortunate.

An important fact apparent from Table 1-4 below is that the major energy-consuming devices tend to be those which intentionally produce heat or cold, such as the air conditioner, clothes dryer, freezer, refrigerator, range and water heater. Devices which drive motors or fans tend to require less power and use less energy than others mentioned above. More economical devices include the washing machine, furnace fan, and vacuum cleaner.

FIGURE 1–3. Use of electricity in the home. The residential sector accounts for about 30% of electric energy used in the United States. Cooking is a major user of electric energy. Using gas is more efficient, but gas may not be available in all locations. Such devices as television and refrigerators are nearly saturated (nearly all homes contain both). The dishwasher is an example of an appliance which will very probably grow significantly in use. Smaller appliances use some power, but it is cooking, heating, and air-conditioning that require major amounts of power in residences. (*Courtesy of the Tennessee Valley Authority.*)

In terms of cost, whether it be dollars or pollution level, energy is the major factor. We see, for example, that it costs far less to run a carving knife than a clock even though the clock is much less powerful. The reason, of course, is that the clock is always "on," whereas the knife is seldom used. The refrigerator uses about the same amount of energy as the range, although it demands far less power. We shall see in Section 1–8 that this condition makes the refrigerator a much more desirable consumer than the range

TABLE 1-4

Typical Household Appliance Power and Energy Requirements

Appliance	Power (Watts)	Annual Energy (KWH)
Air Conditioner (Window)	1,566	1,389
Broiler	1,436	100
Carving Knife	92	8
Clock	2	17
Clothes Dryer	4,856	993
Deep Fat Fryer	1,448	83
Dishwasher	1,201	363
Electric Blanket	177	147
Fan (Circulating)	88	43
Fan (Furnace)	292	394
Food Blender	386	15
Food Freezer (Frostless 15 cu. ft.)	440	1,761
Food Mixer	127	13
Food Waste Disposer	445	30
Frying Pan	1,196	186
Heat Pump (Electric Heating System)	11,848	16,003
Heater (Radiant)	1,322	176
Hot Plate	1,257	90
Iron	1,088	144
Oil Burner or Stoker	266	410
Radio	71	86
Radio-Phonograph	109	109
Range	12,207	1,175
Refrigerator (12 cu. ft.)	241	728
Refrigerator (Frostless 12 cu. ft.)	321	1,277
Refrigerator-Freezer (14 cu. ft.)	326	1,137
Refrigerator-Freezer (Frostless 14 cu. ft.)	615	1,829
Sewing Machine	75	11
Shaver	14	18
Television (BW)	287	362
Television (Color)	332	502
Toaster	1,146	39
Toothbrush	7	5
Vacuum Cleaner	630	46
Washing Machine (Automatic)	512	103
Washing Machine (Non-Automatic)	286	76
Water Heater (Standard)	2,475	4,219
Water Heater (Quick Recovery)	4,474	4,811
Water Pump	460	231

from the standpoint of the electric utility which must provide for both energy and power demand.

Recently a number of groups have proposed that every appliance sold should clearly state the power it requires. From the discussion above it is clear that it may be just as desirable to state the average energy required per year by the device. This information should be of interest to the consumer, since it will provide him with a measure of the cost of operating the device and the relative effect of the device on the environment.

Finally, most of the large energy users on our list provide for basic human needs such as food, heat and cleanliness. Few of the commonly-discussed frills use much energy. To summarize this question of residential usage, we have selected some representative devices to use in a 10,000 KWH/year residence which has an electric water heater and non-electric heating. This summary is given in Table 1-5.

TABLE 1-5.

Residential Use of Electric Power

Use	KWH	Percent
Cleanliness (includes hot water)	5,200	52
Food Storage and Preparation	2,800	28
Lighting	1,000	10
Service Appliances: Iron, Vacuum, etc.	500	5
Entertainment	500	5

Next we ask how industry uses its large share of energy. Table 1-6 gives the energy used by the major industries in this country for 1963. Also shown is the cost of purchased power as a percent of the total product value. The latter figure is interesting because it shows that very little of the cost of manufacturing things is in the electric energy consumed. It is sometimes argued that low-cost electric energy is the essential ingredient in stimulating the growth of an industry in an undeveloped area. The figures in Table 1-6 suggest that it takes much more than the availability of energy to cause industrial growth. This point is discussed in some detail by Sporn.(4)

FIGURE 1—4. Industrial use of electric energy. Industry uses about 43% of all electric energy generated in the United States. In the end, of course, the consumer is the final though indirect user of all electric energy, whether it be industrial, residential, or commercial. As we shall see in Chapter 13, transportation uses 25% of all energy resources in the United States for propelling vehicles. Another 15% is used to build vehicles, roads, and shops, and to provide required transportation sales and maintenance service. (*Courtesy of Pacific Gas and Electric Co.*)

1.6 EXAMPLES OF SOME LARGE-SCALE SYSTEMS

In recent years much work has been done on determining the electric energy required for certain fairly complex or large–scale systems. In this section we discuss two of these systems, the operation of electric mass transit systems, and the operation of sewage treatment plants. In Chapter 14, after we have discussed some of

TABLE 1-6.

Industrial Uses of Electric Energy, 1963

Industry	Electric Energy Consumption: Millions of KWH	Cost of Purchased Power: Percent of Product Value
Chemicals and allied products	109,494	2.0
Primary metal industries	100,264	1.9
Fabricated metal products	9,412	0.6
Paper and allied products	39,572	2.2
Food and kindred products	22,149	0.5
Stone, clay and glass products	18,052	1.5
Transportation equipment	18,898	0.4
Petroleum and coal products	17,781	0.8
Textile mill products	15,427	0.9
Machinery, except electrical	11,730	0.5
Electrical machinery	13,070	0.5
Rubber products	7,637	1.0
Lumber and wood products	6,055	0.9
Printing and publishing	4,099	0.4
Apparel and related products	2,509	0.3
Furniture and fixtures	1,848	0.5
Instruments and related products	2,122	0.5
Leather and leather products	1,019	0.4
Tobacco manufactures	511	0.1
Miscellaneous manufactures	1,880	0.5
Ordnance and accessories	2,091	0.4
ALL MANUFACTURING	405,932	0.86

the technical aspects of generation, we shall consider the interesting case of the proposed widespread use of electric automobiles.

As one would expect, electric mass transit systems require large amounts of electric power and energy. However, this demand may be small compared to the total demand of the supplying system. Also, mass transit systems tend to require much less total energy than private automobiles, though the form of the energy is different. As an example we consider the Bay Area Rapid Transit (BART) system, which began operation in the San Francisco Bay Area in 1972. It is anticipated that the maximum power demand of this system, when it is fully operative, will be about 120,000 KW. This is only about one percent of the total power load of the supplying company, Pacific Gas and Electric Co. With an *annual* growth rate of about seven percent for all purposes, BART's *one-time* growth of one percent should not be particularly significant, considered over the lifetime of the system.

The energy required by BART is estimated at about 0.3 KWH per passenger-mile. This assumes conservatively that 20 percent of the seats are occupied.

Example 1-5

Assume that a BART commuter rides 20 miles each way for each of 225 days of the year. How much electric energy does he consume, and what fraction is this of the average of 7,700 KWH used per year by persons in the United States?

$$E = 2 \times 20 \times 225 \times 0.3 = 2,700 \text{ KWH}$$

$$\frac{2700}{7700} \times 100\% = 35\%$$

Table 1-7 lists a number of particular large industrial users of electric energy, and indicates the approximate percentage of the total U.S. production of electric energy required by each.(5)

TABLE 1-7.

Some Industrial Users of Electric Energy

Industrial Use	Percentage of Total U.S. Electric Energy Production
Steel	4
Aluminum	4
Petroleum Refining	1.8
Metal Can Fabrication	1.5
Motor Vehicles	1.3
Paper Mills	1.2
Plastics and Synthetics	0.9
Weaving Mills	0.7
Household Appliances	0.2
Malt Liquor	0.1
Steam Engines and Turbines	0.04
Electric Lamps	0.04
Costume Jewelry and Notions	0.03

FIGURE 1–5. The Bay Area Rapid Transit (BART) System. When BART started operations in 1972, it was the first totally new rapid transit system in the United States in 50 years. It is only the first of a large number of systems under construction or in planning stages. Much electric energy is needed for such systems. However, this energy is small compared to the total demand for electric energy. Also, mass transit systems are much more energy-efficient than the automobile. (*Courtesy of Pacific Gas and Electric Co.*)

Sewage–treatment energy costs are difficult to state unambiguously because they are highly dependent on plant size, degree of treatment, and on the amount of industry in a specified treatment area. One recent study suggests that a fairly small plant serving about 30,000 people, with an average amount of industry, will require about 65 KWH per person per year for primary and secondary treatment.(6) Hence if such a plant is built in an area, it represents a one–time growth (compared with 7,700 KWH per person per year) of less than one percent.

It has been suggested in recent years that solving environmental problems will require "vast new quantities" of electric energy. Two examples often cited are mass transit and sewage treatment. As we have seen, these systems do *not* appear to require major new-growth *relative* to growth from other causes. This point is developed in more detail in a recent paper by Hirst and Healy.(7)

1.7 CONSERVING ELECTRIC ENERGY

There are two ways to face the crisis of inadequate supply of electric energy to meet demands. One is to increase the supply to meet all demands. The second is to question the need for the demand and seek areas in which demand might be cut. Decreasing the demand, or the rate of increase in demand, can be accomplished by limiting of goods and services, or by increasing the efficiency of energy use.

It would be unfortunate to have to limit the goods or services available to people. Free choice is one of the most desirable qualities of our society. However, when that free choice begins to infringe on other rights or qualities of life, such limitation may become acceptable or even *necessary*, as in the case of a brownout or a blackout. Many people have advocated the changing of electric energy costs to reflect the costs of environmental deterioration, and to require cost of electric energy to increase with the amount used. There is presently (1973) a continuing debate as to the desirability of these approaches. The answer is not clear.

Another approach is to increase the efficiency of systems. Here, it is usually necessary to consider all forms of energy, rather than electric energy alone. For example, it is generally more energy efficient, and less costly, to heat a home by natural gas than by electricity. On the other hand it is much more efficient to transport people by an electric mass transit system than by private gasoline automobiles. Other savings of energy can be accomplished by insulating houses, by using fluorescent instead of incandescent lighting, by recycling aluminum and steel (but probably not broken glass), by using returnable bottles for beverages, and many more.

1.8 THE DEMANDS ON AN ELECTRIC POWER SYSTEM

In previous sections we saw how various sectors of society use electric energy. In this section we consider how society presents its demands for power as a function of time. The demand or use of energy is often called the *load* on the system.

If everyone demanded a constant level of power at all times, the task of meeting this demand would be much simpler than it is. Unfortunately, our demands vary from hour to hour, day to day, and season to season. A curve of typical diurnal (daily) demand in a large metropolitan area is shown in Figure 1-6. There is a minimum power level at all times, even late at night. This minimum demand comes from street-lighting, heating, all-night businesses and factories, etc. About dawn the demand begins to grow as people get up, turn up the heat, cook breakfast, and open offices. There is usually a dip around noon, perhaps 11 A.M. to 2 P.M., and then a second and usually greater peak around 4-6 P.M. as

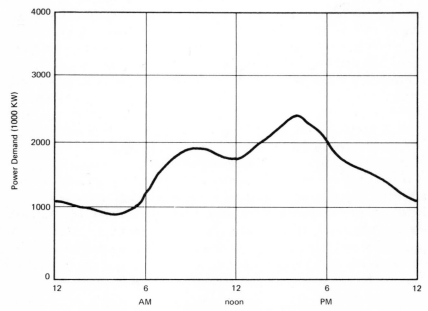

FIGURE 1-6. Typical Daily Power Demand Curve.

people begin to cook evening meals and businesses remain open. After 6 P.M. the peak gradually diminishes into the night.

This same basic demand curve shape applies to all weekdays. On weekends, the curve tends to be generally lower, with less–pronounced peaks. Seasonal variations are also quite important. Figure 1–7 shows the monthly United States energy demand since 1965.(8) Besides showing pronounced seasonal peaks from heating loads in the winter and air–conditioning loads in the summer, the curves also repeat some of the results of Sections 1.5 and 1.6. Both diurnal and annual curves will be affected by such factors as weather conditions and the state of the economy.

It is the responsibility of the power industry to meet the peak power demands of its customers. Since the industry cannot know exactly what demands will occur or when, companies generally try to maintain a fairly comfortable excess of capacity beyond peak demand. This excess is called a "margin." It usually amounts to at least 15 to 20 percent of the peak. This margin allows for excessive or unexpected demands and for breakdowns or "outages." When a company or a region has a low margin and some bad luck with equipment or weather, demand may exceed capacity. The results are blackouts (loss of power to some or all customers) or brownouts (decrease in voltage to customers). The area in and around New York City has experienced this problem during recent summers.

Since it costs money and requires land to build plants to meet power demands, we must have a demand whose peaks are as little above the base load (minimum demand) as possible. Because of the importance of the relation of peak demand to base load, a ratio, called a *load factor*, has been developed to tell us how "peaked" a load demand is. The load factor is the ratio of the average load over a designated period to the peak load occurring in that period.

Example 1–5

Assume that the United States generating capacity has a margin of 25%. Use the figures given in Section 1.3 to find the load factor for the United States in 1971.

Billions of Kilowatt-Hours

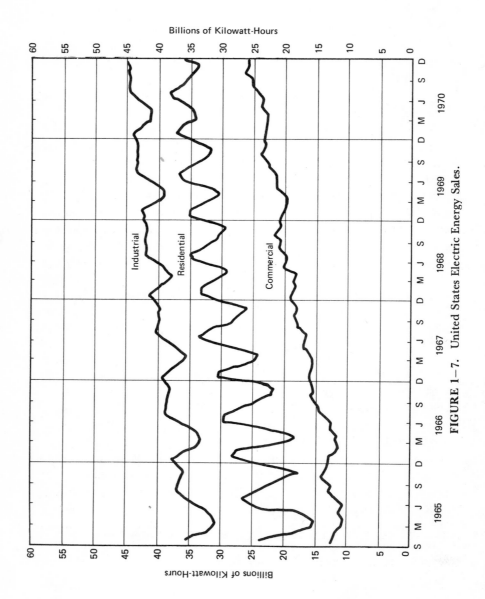

FIGURE 1–7. United States Electric Energy Sales.

Billions of Kilowatt-Hours

22

Since the margin was 25%, the peak power demand was $\frac{360,000}{1.25}$ MW or 288,000 MW. The average load was:

$$\frac{1,600,000,000,000 \text{ KWH}}{24 \times 365 \text{ (hours)}} = 182,000 \text{ MW}.$$

Hence the load factor was $\frac{182,000}{288,000} = 0.63$

✓The nature of the load or demand curve dictates not only the total capacity which a company must install but also the kind of equipment which will be used for generation. We discuss this problem in more detail in a later chapter. At this point, we distinguish between "base load," which is always present, and "peak load," which occurs only at certain times. Base loads tend to be supplied by large fossil-fuel or atomic plants which should run continuously because of the rather long times required to start them. Peak loads tend to be supplied by hydro facilities where available; or increasingly by pumped-storage plants; or by gas turbines, smaller fossil-fuel plants (often older units), or by internal-combustion engines.

REFERENCES FOR CHAPTER 1

1. "A Review and Comparison of Selected United States Energy Forecasts," Prepared for the Executive Office of the President by Pacific Northwest Laboratories of Battelle Memorial Institute, U.S. Government Printing Office, December, 1969.
2. "Questions and Answers about the Electric Utility Industry—1970-71 Edition," Edison Electric Institute, New York, 1970.
3. Edison Electric Institute, Marketing Division, New York, 1969.
4. P. Sporn, *Technology, Engineering, and Economics,* MIT Press, Cambridge, Mass. 1969.
5. "1967 Census of Manufacturers—Fuel and Electric Energy Consumption," U.S. Dept. of Commerce MC67 (S)-4 U.S. Government Printing Office.
6. C.A. Washburn, "Clean Water and Power," *Environment,* Vol. 14, No. 7, September, 1972, pp. 40-44.

7. E. Hirst and T. Healy, "Electric Energy Requirements for Environmental Protection," Conference on Energy: Demand, Conservation and Institutional Planning, Massachusetts Institute of Technology, Cambridge, Mass. February 12–19, 1973.
8. Federal Power Commission, "Electric Power Statistics," November, 1970.

ADDITIONAL READING FOR CHAPTER 1

1. E. Vennard, *The Electric Power Business*, McGraw–Hill, New York, 1970.

   This is a very readable overview of the electric power industry from the viewpoint of the industry. It discusses structure of the industry, growth, regulation, loads, selling, pricing, and community affairs.

2. D. Meadows, *et. al.*, *The Limits to Growth*, Universe Books, New York, 1972.

   This is perhaps the most widely read and controversial book on growth in our time. It hypothesizes some growth models and uses computers to project possible future states of mankind. Most of the prospects are bleak. The book has been praised and damned. It should be read, but critically, for its strengths and its weaknesses.

3. H. Landsberg and S. Schurr, *Energy in the United States: Sources, Uses, and Policy Issues,* Random House, New York, 1960.

   This is a general review of the energy industry. It is a little outdated, but much of the information is still valid. The book covers much of what is in this chapter, though in more detail.

4. *Patterns of Energy Consumption in the United States,* Stanford Research Institute, Office of Science and Technology, U.S. Government Printing Office, January, 1972.

   This is a recent comprehensive government report on how energy in all forms is used by different sectors of society. It has about 200 pages of important data on energy use. It is not light reading.

5. O.L. Culberson, *The Consumption of Electricity in the United States,* Oak Ridge National Laboratory, ORNL-NSF-EP-5, Oak Ridge, Tenn., June, 1971.

This is a good general study of how electric energy is used. It is interesting and easily read.

6. J. Holdren and P. Herrera, *Energy,* Sierra Club, San Francisco, 1971.

This book is divided into two parts. The first is an excellent general review of the energy question. The coverage is similar to that in this textbook. The book is non-mathematical and easily read. The second part of the book is a series of very interesting case studies of proposed power systems. Taken as a whole, *Energy* is a very fine reference for students of the subject.

PROBLEMS FOR CHAPTER 1

General Problems

1.1. Let us assume that the total power of the light bulbs in your home is 2000 watts and you operate them an average of 5 hours per day. If electric energy costs you $.024/KWH find the daily cost of lighting your home. ($.24)

1.2. Assume that the population growth rate were reduced to zero and the per capita energy–use growth rate remained at about 5.5%. What is the doubling period? That is, in how many years will energy use double? (13 years)

1.3. What percentage of the time are clothes dryers in operation? (See Table 1-4.) (2.33%)

1.4. Estimate the amount of electric energy used in your home each year and the cost of that energy.

1.5. How does the cost of operating a device compare with the cost of buying the device? To find out, select an appliance, estimate its cost and its lifetime, average the cost over the lifetime to get an approximate cost per year, and compare this with the operating cost per year. Assume an energy cost of $.025/KWH. Try this for a clothes dryer, a television, a shaver, and others. Any observations?

1.6. The "operation factor" is defined as the ratio of the total time of actual service, of a machine or equipment, to the total period of time considered. What is the operation factor of a a) clothes dryer; b) furnace fan; c) 12 cu. ft. refrigerator; d) clock? (0.023; 0.154; 0.345; 1.000)

1.7. Find the approximate energy used per day corresponding to the demand curve of Figure 1-6. Do this by considering the demand constant over a one–hour period and finding the energy for each of the 24 hours during the day. Those familiar with calculus may wish to think in terms of using the trapezoidal integration rule. Also, find the approximate load factor. (37,000,000 KWH; 0.67)

Advanced Mathematical Problems

1.8. Use a mathematical proof by induction to show that Equation 1-1 is valid. This requires that you: 1) show that it is valid for n = 1; and 2) assume it is valid for n = k and show that it is then valid for n = k + 1. (Hint: To see how you get from n = k to n = k + 1, recall that the increase in energy is just r times the energy in year k.)

1.9. Look up the expression "geometric series" in a mathematics text, and relate it to the problem of exponential growth.

1.10. The value of differential calculus is that it allows us to study phenomena which change over differentially small periods of time. The student of calculus can study exponential growth through the simple differential equation $\frac{dE}{dt} = kE$ which says that the instantaneous rate of change of energy is directly proportional to the present level of energy. Show that

$$E = E_o e^{kt}$$

is a solution to the above differential equation by substitution of this proposed solution into the differential equation.

1.11. Compare the solution to the differential equation given in Problem 1.10 with Equation 1-1. Let one year be represented by a time $t = T$. Find the relation between k and r. If $r = 0.07$, what is k? ($k = 0.0677$/year)

1.12. Suppose that the population of the United States continues to grow at 1.5% per year. What will the population be in a) 20 years; b) 100 years; c) 200 years; d) 500 years? Assume it is 205 million now. (275 million; 910 million; 4 billion; 350 billion)

1.13. Suppose that energy–use growth rate per capita is 5% per year. Draw a curve of the growth assuming an initial energy E_0 if the population growth rate is a) 0%; b) 2%. Consider a 20 year growth interval.

1.14. Plot the number of years required to double energy use as a function of the yearly percentage growth rate.

Advanced Study Problems

1.15. Find the power required to run as many devices, appliances, etc. as you can, and make a list of them. You can usually find this data stamped on a nameplate somewhere on the device. Compare this with the list given in this chapter. Why should there be any differences?

1.16. Read *The Limits to Growth* (see list of books), and then obtain three or four critical reviews of the book. They appeared widely in sources such as *The New York Times* (excellent review), *Saturday Review, Time,* and many more. After reading the book and a number of reviews, point out the areas of strengths and weaknesses of the book. Will this book "last"? How will history regard it?

1.17. Why is energy of value to man? Outline its history of use. What would happen to our present society if a) energy growth was stopped so that the amount used became constant; b) energy use was halved; c) energy use was cut to zero?

1.18. If energy is a *sine qua non* for industrial development, is it also sufficient for development? See P. Sporn (Reference 4.). Explain!

2

ELECTRIC ENERGY FROM OTHER FORMS

Nature hath no goal though she hath law.

John Donne

How does nature store energy? How can energy become available to man? How can energy in non–electrical forms be converted into electrical energy? These are the central questions of this chapter, and they are a major focus of the entire book. We also consider in this chapter the operation of the electric generator, the last step in a sequence of operations which produces electric energy. Finally, we briefly consider two fundamental energy laws which apply to all conversion processes, and which have profound implications for all of our studies.

The discussion in this chapter is rather general. The operation of and problems relating to specific generation schemes are discussed in detail in later chapters.

2.1 STORAGE OF ENERGY IN NATURE

As we saw in Chapter 1, energy is the capacity to do work. The central focus of man's technological growth has been the finding of ways to use energy, which nature stores in many different forms. The ways in which nature stores energy are shown in Table 2-1. The forms listed in the column on the left are those which either are presently converted to electric energy, or could be, if they were developed and found economically satisfactory. The type of electric–energy generation plant associated with each form is shown on the right. The numbers in parentheses are the approximate percentages of energy contributed by each form of conversion in the United States in 1970. On the bottom of the table are shown the remaining forms of energy, which do not presently appear to offer prospects for electric energy generation.

TABLE 2-1.

Forms of Natural Energy

Form of Energy	Associated Electric Energy Generation Scheme
Gravitational Potential	Hydroelectric (16%) Tidal*

Chemical	Fossil fuel to steam (82%)
	   coal (46%)
	   oil (12%)
	   gas (24%)
	 Wood and other burnable fuels
	 Magnetohydrodynamic (MHD)
	 Fuel cells

Nuclear	Fission (2%) Fusion
Heat	Geothermal (0.05%)
Kinetic	Windmills
Wave	Wave–driven (ocean or lake)
Radiation	Solar

Other forms of energy: Strain (spring); spin (rotational kinetic); latent heat (evaporation and melting); electromagnetic–wave; electric; magnetic.

*Where no percentage is indicated, the amount of energy generated is insignificant at this time.

We have categorized generation plants according to the form in which nature most immediately presents the energy rather than according to the originating or initial form. For example, hydro-electric power is classified as originating from a gravitational form even though the energy which lifts water into mountain streams from the oceans comes from the sun.

We turn now to a brief discussion of some of the most important of these energy forms.

Potential energy is energy which a body has solely by virtue of its position in space. More particularly, gravitational potential energy is energy which a body has by virtue of its position in a space dominated by a gravitational field, such as the earth's. Consider Figure 2-1. The rock at the top of the hill (point A) has a potential energy *relative* to points B, C and D. (We cannot associate a potential energy with the rock except in relation to some point in space.) The rock has the same potential energy with respect to points B and C because these points are at the same elevation. The

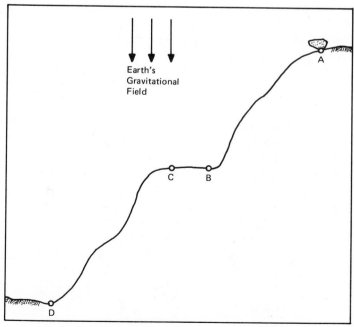

FIGURE 2–1. Gravitational Potential Energy.

distance in the *direction* of the earth's gravitational from A to B
or C is the same. But the distance *down* to D is greater and therefore the rock has a greater gravitational potential energy with respect to D than with respect to B or C.

The above statements should be intuitively appealing. Suppose
the rock rolled down the hill. We know that as it rolled it would
gain speed and would be capable of doing more work (such as
smashing other rocks or denting the earth) at point D than at
points B or C. We also know that the rock's capacity to do work
will depend on its weight. The gravitational potential energy of a
body of weight W located at an elevation H greater than some
point will have a potential energy with respect to the reference
point.

$$P.E. = WH \qquad\qquad (2\text{-}1)$$

We make use of this form of energy in hydroelectric power plants
by storing rain and snow water behind dams, and then allowing
it to fall through hydraulic turbines to generate electric energy. We
explore potential energy and its application to hydroelectric plants
in Chapter 3.

Most substances, including those which we use as fuels, in fossil–
fuel power plants, can be said to store *chemical energy*. They
store energy in the sense that when they enter into a chemical reaction with oxygen to form a new chemical compound, they release energy as heat. For example, when carbon combines with
oxygen to form carbon dioxide, two molecules come together because of atomic forces of attraction, causing a new chemical bond.
The result is that the new molecule of carbon monoxide has some
additional energy which appears as heat. Coal, oil, and natural gas
are commonly used as sources of chemical energy or fuels in electric power plants. (Because of their development over millions of
years as fossilized vegetation, oil, gas and coal are called *fossil
fuels*.) The way in which the heat from fossil fuels is used to generate electricity will be explained briefly in the next section and in
detail in Chapter 5.

Nuclear energy is associated with the mass of the nuclei of
atoms. A part of this energy is released when nuclei disintegrate

during an atomic reaction. There is a loss of mass when nuclei break up into new components, and a corresponding gain in energy equal to $E = \Delta mc^2$ where Δm is the change of mass and c is the velocity of light. The energy released in a nuclear chain reaction can then be used in essentially the same way as the energy released from burning fuel. We shall consider nuclear power plants in detail in Chapter 6.

2.2 ENERGY CONVERSION SCHEMES

At the present time there are two important large–scale electric energy generation schemes. Both have as their final stage a rotating machine called an *electric generator* which produces the desired electric energy. The two schemes are sketched in Figure 2–2.

The first scheme, shown in Figure 2–2a, uses water under pressure (from gravitational potential energy) to drive a hydraulic turbine (a special form of waterwheel). The shaft of the turbine is physically connected to the shaft of the electric generator, causing the generator to rotate. The rotating generator produces electricity in a way to be described in the next section. This scheme is entirely mechanical, and it can be quite efficient. As much as 80 percent to 90 percent of the energy available from the water can be converted to electric energy.

The second scheme is similar to the first but it has one additional element, a steam generator, which drives a steam turbine rather than a hydraulic turbine. (To the right of the turbines in Figure 2–2 the schemes are identical.) The steam generator converts the chemical or nuclear energy of the fuel into thermal energy in the steam which runs the steam turbine. Unfortunately, heat processes are inherently inefficient, as we shall see in Chapter 5. The result is that the second scheme is much less efficient than the first. Typical efficiencies are 30–40 percent, with some very efficient modern units exceeding 40 percent. One might ask why we use the second scheme if the first is so much more efficient. The answer is that relatively little water power is available, and all fuel–burning systems require the heat process. As we shall

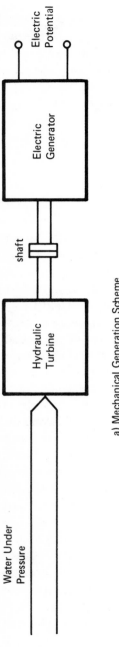

a) Mechanical Generation Scheme

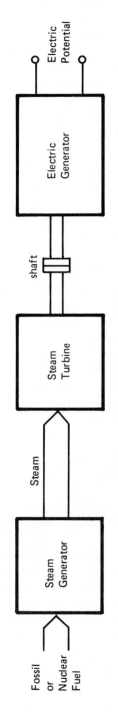

b) Thermal-Mechanical Generation Scheme

FIGURE 2–2. Major Present Electric Energy Generation Schemes.

see in Chapter 2, hydroelectric plants require a combination of desirable geography and weather conditions. Potential locations for
hydro dams and reservoirs are often desired for other purposes.
Finally, hydro projects often require high initial capital costs which
cannot be justified.

Some of the proposed future generation methods use schemes
different from the two in Figure 2-2. For example, one fusion
generation process avoids both turbine and rotating generator. We
discuss this scheme and some others in later chapters.

2.3 THE ELECTRIC GENERATOR

In a series of very famous experiments conducted in the fall of
1831, Michael Faraday showed that when a magnetic field in the
vicinity of a loop of wire is suddenly changed, a voltage is induced
in the wire. Voltage is a measure of the potential *energy* of a unit
of electric charge with respect to some point. It gives the capa-
bility of an electric current to transfer energy from one point to
another. This fundamental physical phenomenon is produced in
the electric generator, which is then able to cause electric energy
to flow through wires to a desired place of use. The device or system which finally uses the electric energy is called a *load*.

With Faraday's basic observation, it remained only to find a configuration for a practical working generator. Consider Figure 2-3a.
We can produce a voltage V by moving a magnet which is close to a
loop of wire. This effect, however, will be small and difficult to
repeat. To increase the magnitude of the effect we increase the
number of turns of wire which the magnetic field affects. To facili-
tate the task of moving the magnet with respect to the coil, and to
make this movement uniform, we place the magnet on a rotating
shaft. These two improvements are shown in Figure 2-3b.

The simple generator of Figure 2-3b has all of the basic features of modern generators. However, today's generators have a
number of refinements. First, permanent magnets are not strong
enough to produce the desired rotating magnetic field. For this
reason wire is wound on the rotating bar and electric current is

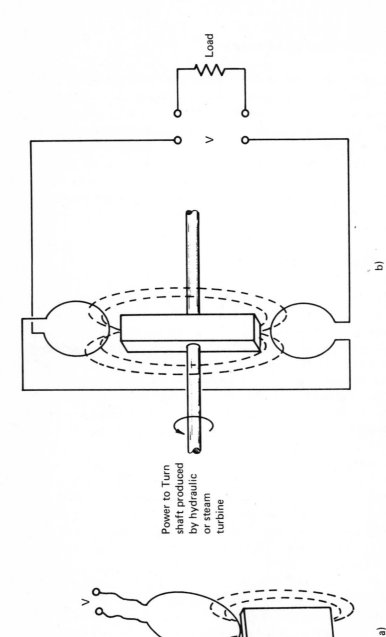

Load

V

Power to Turn
shaft produced
by hydraulic
or steam
turbine

b)

a)

V

FIGURE 2–3. Basic Electric Generators.

fed to the wire through sliding contacts, called slip rings. This forms a so-called electromagnet, which serves the same role as the permanent magnet.

Second, rather than put only two coils around the stationary part of the generator, we actually put many. This improves space utilization and gives flexibility in selecting the speed of rotation of the generator for a given generation frequency.

Let us turn our attention briefly now to the voltage produced by the generator. The voltage produced by almost all large power generators at this time is an AC (alternating current) voltage, which changes from positive to negative once every half cycle. This voltage form is shown in Figure 2-4. It is a so-called "sinusoidal" voltage, with the form:

$$V = V_o \sin 2\pi f t \tag{2-2}$$

where V_o is its maximum, f is the frequency in cycles per second (or Hertz, abbreviated Hz) and t is time.

The standard frequency in the United States is 60 Hz. In many parts of the world it is 50 Hz, and some other frequencies are also used. The rated voltage of a generator is usually given not as V_o but as $V_o/\sqrt{2}$, and this value is typically 10,000 to 15,000 volts, for a large generator.

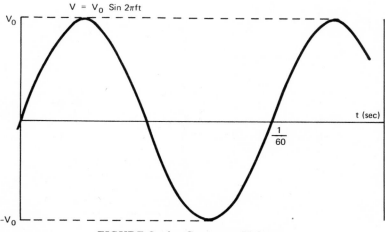

FIGURE 2—4. Generator Voltage

It is possible to generate DC (direct current) voltage. For many years, particularly at the end of the nineteenth century, there was much debate over the relative advantages of each system. One of the most important advantages of AC voltage is that its level can be increased or decreased by any desired factor by a transformer, such as is sketched in Figure 2-5. This is a device with two coils which are coupled by a magnetic field. The ratio of voltage levels is approximately equal to the ratio of the number of turns. Transformers are used to increase the generator voltage from around 12,000 V to as much as 500,000 V or more when it is necessary to transmit energy over long distances. This is done because transmission at high voltages, for a fixed amount of power, is much less wasteful or inefficient in terms of power losses in the transmission lines.

On the negative side of the picture, high voltage transmission lines require large towers because of the necessary insulation and separation of such lines. Large towers are not particularly attractive, and opposition to use of land for power transmission towers and lines is becoming an increasingly important part of the siting problem; that is, the problem of where to locate the power plant.

Let us consider now what happens when a load, or energy demand, is placed on a generator, and energy is taken from the gen-

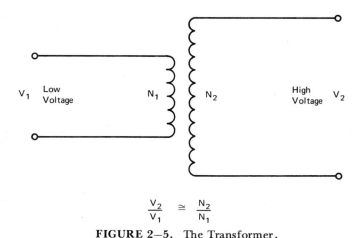

$$\frac{V_2}{V_1} \cong \frac{N_2}{N_1}$$

FIGURE 2—5. The Transformer.

erator. Consider the sketch in Figure 2-6. When the generator terminals are open (that is, no load is connected) the only mechanical power the shaft must provide is that needed to overcome friction and magnetic losses. Assume now that a load, such as a toaster, is connected to the generator. (It is of course necessary that the load be designed to work at the generator's voltage. Otherwise we assume that a voltage step–down transformer is placed between generator and load.) When the load is connected it draws a current from the generator. The current's moving electrons develop a magnetic field. A force exists between this magnetic field and that of the electromagnet which tends to keep the electromagnet and hence the shaft from turning. To overcome this force more mechanical power must be applied to the shaft. If additional load, such as an iron, is added, the current goes up, the force goes up, and hence the shaft power goes up. In this way the electric power required for various loads is reflected as required mechanical power on the generator–turbine shaft.

2.4 TWO FUNDAMENTAL LAWS

We close this chapter on energy conversion with a brief consideration of two extremely important statements, called the First and Second Laws of Thermodynamics. There are many ways to express these laws. The following expressions are useful for our viewpoint.

1) **The First Law of Thermodynamics:** Whenever a process undergoes a cyclic change, energy is always conserved.
2) **The Second Law of Thermodynamics:** It is impossible to construct an engine which, when operating through a complete cycle, will convert all of the heat supplied to it into work.

Neither of these laws is obvious; neither can be proved. They are based on observations. There is no confirmed evidence that either has ever been violated. We accept them because they have proved thus far to be true.

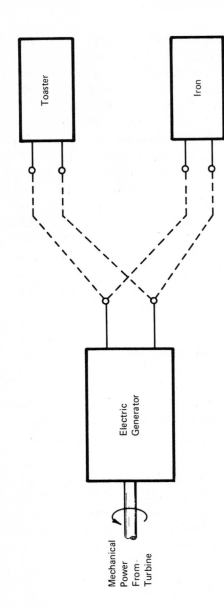

FIGURE 2-6. Loading the Generator.

The First Law of Thermodynamics is sometimes called the Law of Conservation of Energy. It tells us that energy can be neither created nor destroyed. It can only be transformed from one form to one or more other forms. All of the energy we introduce into a system must go somewhere. Suppose that we decide to burn a certain amount of coal to produce electric energy. Some energy will remain as unburned coal particles, some will be lost as heat in the stacks, or dissipated as heat radiated from the boiler or from steam pipes. Some energy is lost (due to the Second Law) as waste heat; some goes into friction losses, and some (about 40 percent) is converted to electric energy. All of these energies must add up to the total initial energy released in the coal combustion process. This is summarized in Figure 2-7, where a thermal electric steam plant is used as an example of a system.

The implication for all of our electric energy generation schemes is that we cannot get electric energy out unless we put some kind of energy in, or that if we put in a certain amount of energy, all of it must be accounted for. Since our systems have inherent inefficiencies, we conclude that *every generation scheme will produce waste heat.* The percentage of waste heat in the case of thermal plants is quite large and poses major engineering and environmental problems. We discuss these in some detail in Chapters 5 and 12.

The Second Law of Thermodynamics tells us that we cannot transform all of the heat input to an engine into mechanical work. Some heat must be wasted. There are many forms of the Second Law, but this one seems particularly relevant to our studies because of its emphasis on waste heat. This waste heat, which is the major energy output shown in Figure 2-7, is inevitable in any heat–work conversion process. In Chapter 5 we pursue this question quantitatively to see just how much waste is inevitable.

ADDITIONAL READING FOR CHAPTER 2

1. B. Chalmers, *Energy*, Academic Press, New York, 1963.

     This is a general textbook in energy, written at an introductory level. Most of it can be understood with little or no

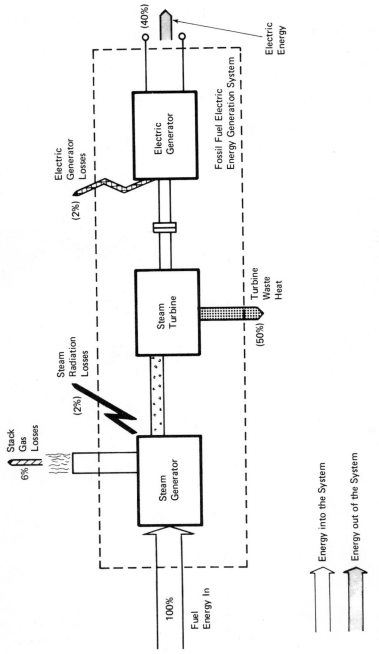

FIGURE 2–7. The First Law of Thermodynamics.

(40%)

Electric Energy

Electric Generator Losses

(2%)

Electric Generator

Fossil Fuel Electric Energy Generation System

Steam Radiation Losses

(2%)

Steam Turbine

Turbine Waste Heat

(50%)

Stack Gas Losses

6%

Steam Generator

100%

Fuel Energy In

Energy into the System

Energy out of the System

background in physics or mathematics. Included are extensive sections on most of the material discussed in Chapters 1 and 2 of this book, as well as some of the later chapters.

2. H.C. Van Ness, *Understanding Thermodynamics*, McGraw-Hill, New York, 1969.

This short paperback should be very interesting to the student with some background in physics and/or thermodynamics. It can be read, in part, by students without this background. It has a particularly good discussion of the Second Law of Thermodynamics. It also has a short discussion of thermal electric power plants, which we shall be studying in Chapters 5 and 6.

3. H.T. Odum, *Environment, Power, and Society*, John Wiley and Sons, Inc., New York, 1971.

This book deals with most of what is in this chapter, but goes well beyond it to a complex and sometimes esoteric discussion of man's relation to nature, to energy, and to the biological community. It includes chapters on politics and religion. Part or all of it may be read with profit by more serious students.

PROBLEMS FOR CHAPTER 2

General Problems

2.1. Consider a windmill used to mill flour. Discuss the system in terms of the energy in and the energy out through various paths.

2.2. Consider a sky-diver who parachutes from an airplane. Discuss how energy is stored when the diver is in the plane and in what way this energy is eventually dissipated.

2.3. Explain why the following question cannot be answered. What is the potential energy of a one-pound block of wood?

2.4. Consider an electric fan as an energy system and discuss the energies into and out of the system.

Advanced Mathematical Problems

2.5. Show that V, as defined by Equation 2-2, can be expressed as the imaginary part of a rotating vector, or phasor,

$V_o \, e^{j2\pi ft}$. How does this relate to the generation of the voltage V by a rotating magnetic field?

2.6. Find the potential energy, with respect to a given plane, of 3 cubic feet of steel, lifted 10 feet above the plane. If the steel is lifted in 4 seconds, what is the average power expressed in units of pound-feet/sec? (Hint: the data required to solve this problem is not given in this book.)

2.7. A generator has an output voltage of 13,000 volts. It is desired to transmit power over a 500,000 volt transmission line. What is the necessary transformer–turns ratio (N_2/N_1)?

Advanced Study Problems

2.8. Perform some real experiment in which energy is stored in some form and then converted to another form. Describe the *actual experiment* which you perform, discussing energy forms, conversion processes, etc.

2.9. Look up the Second Law of Thermodynamics in a number of books in physics, thermodynamics, chemistry, etc. Give a number of forms of the Law, and explain why different forms are often of interest to different people.

2.10. Build and test a small electric generator.

2.11. Read Chapter 8 in Reference 3 and discuss critically the "Ten Commandments of the Energy Ethic."

3

HYDROELECTRIC POWER PLANTS

The water before, and the water after, now and forever flowing, follow each other.

> The Zenrin
> (Edited by R. H. Blyth)

Strangely enough, it all starts with a thermonuclear explosion on the sun, some 93 million miles away. Energy is radiated from the sun into space. Some of it reaches the earth. The atmosphere is warmed and water evaporates from the ocean. A small low pressure cell develops far out in the Pacific Ocean. A storm is born; it grows as it moves. It accumulates great quantities of moisture by the time it reaches the Pacific Coast of the United States. The water cycle, discussed in Section 3.1, has begun. The rain and snow which are to fall are nature's gift to man to use as he will.

In this chapter we consider the use of stored water to generate √ electric energy. Man has been using water from rain and snow since the beginning of his existence on earth. In the early history of man, water from the rain cycle was used for drinking, irrigation, cleaning, transportation, etc. In more recent centuries man has harnessed the energy in flowing water for a variety of purposes, including grinding flour, lifting water for irrigation, and powering forges and sawmills. It is not surprising that much of man's early supply of electric energy came from stored water. Today, hydroelectric power is an important part of the electric energy generation picture, contributing about 16 percent of our present electric energy needs.

Hydroelectric power has a number of advantages which make it very desirable in some situations. Among the advantages are the following:

1) Costs of operation and maintenance are low.
2) Generating plants have a long life.
3) Unscheduled breakdowns are relatively infrequent and short in duration, because the equipment is relatively simple.
4) Hydroelectric turbine–generators can be started and put "on–line" very rapidly.
5) Hydroelectric facilities have a relatively small impact on the environment in some cases.

We turn now to the question of how water is used to generate electric energy.

3.1 THE WATER CYCLE

The water cycle is one of the most important phenomena in nature. It is the process by which water evaporates, primarily from the oceans, into the sky, condenses, and eventually falls back to earth as rain or snow. It goes on continuously and it is a critical element of man's life on earth.

The major steps in the water cycle are shown in Figure 3–1. The primary source of energy is the sun, which starts the cycle by heating the ocean. The heated water partially evaporates; that is, it releases some water vapor to the air. (We shall restrict our dis-

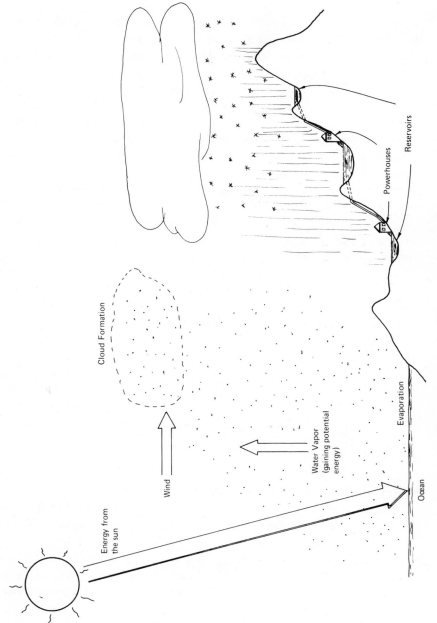

FIGURE 3–1. The Water Cycle and Hydroelectric Power.

47

cussion of the water cycle to the ocean as a source of water vapor. Actually, water evaporates from land, vegetation and even animals, and this can be significant in some regions. However, the oceans are the major source of water vapor.) The water vapor released from the ocean is lighter than air and rises. It is moved as it rises by winds, whose energy also originates from the sun. We are interested here in water vapor, which is caused to drift over land masses, as suggested by the picture. This water vapor may form clouds. If there is sufficient water vapor available, and, if other necessary meteorological conditions are met, the water vapor condenses and falls as rain or snow. A particularly marked effect occurs if the clouds are blown against a range of hills or mountains. In this case the clouds are forced to rise and the resulting cooling increases the amount of precipitation. As the cloud passes over the mountain it tends to fall again, warm up and actually absorb moisture from the air and ground rather than produce precipitation. Hence the windward sides of mountain ranges tend to get much more rain than the opposite sides. This is the so–called rain shadow effect; it is found in many parts of the world.

Water which falls as rain or snow either evaporates again, enters the ground to irrigate vegetation, or flows in streams into lakes and eventually back to the ocean. This completes the water cycle. It is this flowing water, at some elevation higher than the ocean, which can be harnessed to produce hydroelectric power. Let us see how the water cycle provides us with this great source of energy.

3.2 THE POTENTIAL ENERGY OF WATER

As we saw in Chapter 2, the potential energy (P.E.) of a body is its capacity to do work by virtue of its position. Water at some elevation greater than a reference plane (such as the ocean) has a capacity to do work because of the earth's gravitational force, which will cause the water to fall when it is released. Consider the diagram in Figure 3–2.

In Figure 3–2a an object with weight W is at a height H above

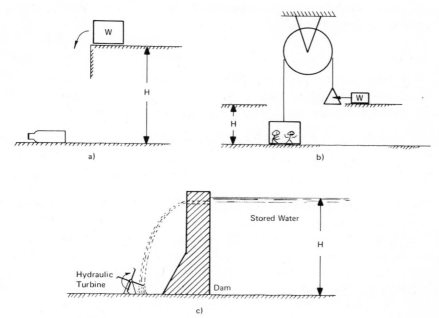

FIGURE 3–2. Using potential energy to do work.

a reference plane. If the object is pushed over the edge, it will fall to the reference plane and in so doing change its potential energy into kinetic energy, which is energy of motion. (Every moving body has kinetic energy.) A little of this energy will heat up the air, but most of it will be converted into heating the object and the reference plane when the object strikes the plane. Now let us put a bottle on the plane. Will the object break the bottle? We are really asking whether the object has enough *energy* to break the bottle when it reaches it. Intuitively we sense that the answer is yes if, for a constant height, the weight of the object is great enough; or, for a constant weight, the height is great enough. And it turns out that the gravitational potential energy of an object is defined as just the product of its weight and its height.

P.E. = WH   (3–1)

If the weight W is expressed in pounds, and the height H is in feet, the units of P.E. are foot–pounds (ft.–lbs.). (In a hydroelectric project, H is usually called *head*.)

Example 3-1

a) What is the potential energy of 100 lbs. of water 500 ft. above the ground? b) How many lbs. of water would have to be placed 25 ft. above the ground to obtain the same P.E.?

a) P.E. $= 100 \times 500 = 5 \times 10^4$ ft.-lbs.

b) $W = \dfrac{P.E.}{H} = \dfrac{5 \times 10^4}{25} = 2 \times 10^3$ lbs. $= 1$ ton

Clearly, if the height of the water decreases, we will need more water to obtain the same energy.

If you wish to do more work you must increase either the weight or the height of the weight above a reference. We will see that in the case of hydroelectric power, either of the two factors may be strongly emphasized.

As an example, Figure 3–2b suggests a simple way in which potential energy can be used to do work. An elevator rests on some plane. One way to raise the elevator is to move a weight W (heavier than the loaded elevator) into a platform attached to the elevator through a rope and pulley, as shown. The weight will move down and the elevator will move up. As always, energy is conserved. The potential energy of the weight W is changed into potential energy of the elevator and its load in its raised position, losses in the pulley system, in air resistance, and in the braking system required to slow and stop the moving bodies at the new steady–state position. (A system is said to be in "steady–state" whenever some particular variables of interest, such as position of the weight W and the elevator in this case, cease to change.)

Finally, in Figure 3–2c we come to the example of basic interest. An obstruction or dam impedes the flow of rainwater toward the ocean, storing it in a reservoir. Water is then released from a height H and allowed to fall on a water–wheel, or hydraulic turbine, which turns on its shaft. The shaft is coupled to an electric generator producing electric energy as the process was described in Chapter 2.

3.3 HYDROELECTRIC POWER

Man is usually concerned not only with capacity to do work but also with the rate at which work is done. As an example, let us return to Figure 3-2b. The occupants of the elevator want to reach the new level, representing a change in height H, but they also want this change to take place fairly rapidly. They probably would not be happy to have a very slow rate of ascent. The *time rate* at which work is done, or energy is expended, is of great interest. This is true in most situations.

The time rate of doing work is called *power*. Power is measured by units of foot-pounds per second. If we look at Figure 3-2c again, it is evident that the power depends not only on the weight W of the water released and the height H from which it is released, but also on the *rate* at which it is released. With the definition above we can easily determine the power of our hydroelectric plant. Assume that some weight of water W falls the distance H in the time T. Then the power is:

$$P_W = \frac{\Delta P.E.}{T} = \frac{HW}{T} \qquad (3-2)$$

where $\Delta P.E.$ stands for the change in potential energy as the weight W falls a distance H.

The weight of water can be expressed as the weight density w (pounds per cubic feet) times the volume of water V (cubic feet).

$$P_W = \frac{HwV}{T} = Hw\frac{V}{T} = HwQ \text{ (foot-pounds per second)} \qquad (3-3)$$

where $Q = \frac{V}{T}$ is the volumetric rate at which water flows out of the reservoir, expressed in cubic feet per second (cfs). Power as defined in Equation 3-3 is sometimes called water power, as it is the total power the falling water contributes to the system. However, a certain amount of this power is lost due to friction in the tubes which carry the water, incomplete transfer of all of the water's energy to the hydraulic turbine blades, and other factors. To account for these losses we multiply P_W in Equation 3-3 by e_t, a number less than 1, called the *turbine efficiency*, to obtain the turbine power. That is:

(turbine power) = (water power) $\times$ e_t

The turbine efficiency is a complex function of the type of turbine used, the head, the percentage of the maximum power being generated, and other factors. It is usually in the range of 80–95 per cent, with a common value in the mid–eighties.

Ordinarily we express the turbine power in horsepower, rather than foot pounds per second. Since 550 foot pounds per second equals one horsepower and water weighs 62.4 pounds per cubic foot, turbine power becomes:

$$P_t = \frac{e_t HQ \times 62.4}{550}$$

$$= \frac{e_t HQ}{8.8} \ (hp) \tag{3-4}$$

Example 3-2

Hydraulic turbines at the San Luis Pumping–Generating plant in central California have a water flow rate of 2,130 cfs when the head is 327 ft. The efficiency for these conditions is about 92 per cent. Find the turbine horsepower.

$$P_t = \frac{0.92 \times 327 \times 2,130}{8.8} = 73,000 \ hp$$

The shaft of the turbine is connected to a generator. Since the generator has some losses also, the electric power out of the generator is obtained by multiplying the turbine power by another number less than 1, called e_g, the generator efficiency.

(generator power) = (turbine power) $\times$ e_g
$\qquad\qquad\qquad$ = (water power) $\times$ e_t $\times$ e_g
$\qquad\qquad\qquad$ = (water power) $\times$ e

where e = $e_t e_g$ is the overall efficiency of the water tunnel–turbine–generator system. Note that the overall efficiency is the product of the two efficiencies of operations which are in "series" or "cascade." These terms indicate that the operations or processes occur sequentially, with the second operation using the output of the first operation. It is always true that when operations occur in this cascade form, the total efficiency is the product of the efficiencies of all the cascaded operations.

The generator power, which is electric power, is expressed in kilowatts (KW). There are 0.746 kilowatts in one horsepower. Using this relation, the new expression for total efficiency, and Equation 3-4, we obtain, for generator power:

$$P_g = \frac{eHQ}{11.8} \quad (KW) \qquad\qquad (3\text{-}5)$$

The efficiency of a modern high-power electric generator is very

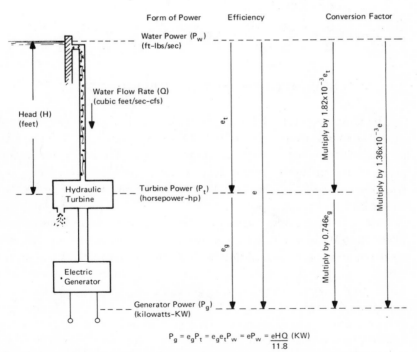

$$P_g = e_g P_t = e_g e_t P_w = e P_w = \frac{eHQ}{11.8} \ (KW)$$

FIGURE 3—3. Summary of Hydroelectric Power Relations.

high, perhaps near 98 per cent. Hence, e is nearly equal to e_t; it has an approximate range of 80 to 95 per cent.

The points made in this section are summarized in Figure 3–3.

Example 3–3

We have a reservoir which can store up to 4.35×10^9 cu. ft. of water (100,000 acre–ft.). The head is 300 ft. How long will we be able to generate 500,000 KW before we empty the reservoir? Assume that the head does not change. Total efficiency is 85 per cent.

$$Q = \frac{P_g \times 11.8}{eH}$$

$$= \frac{5 \times 10^5 \times 11.8}{0.85 \times 300}$$

$$= 23,100 \text{ cfs}$$

$$\text{Time} = \frac{V}{Q} = \frac{4.35 \times 10^9}{2.31 \times 10^4} = 1.98 \times 10^5 \text{ sec.} = 2.29 \text{ days}$$

3.4 TYPES OF HYDROELECTRIC POWER FACILITIES

The simple sketch of Figure 3–2c shows all of the basic elements of a hydroelectric power facility. Likewise, Equation 3–5 gives the power relation common to all such facilities. However, there are a great many different forms which these facilities may take. It is the objective of this section to explore some ways of generating electricity from water power.

One of the simplest and probably earliest examples of man's harnessing of water power was the use of flowing water in a river to drive some sort of paddlewheel. Such a scheme can be adequate for pumping water, grinding flour, and performing certain other tasks. In theory it can be used to generate electric energy. However, variability in stream flow will lead to variations in the electric generator voltage and frequency. One of the purposes of a dam is to smooth out such variations by storing water and releasing it at a controlled rate. (See Figure 3–4.)

FIGURE 3–4. Morrow Point Dam, Colorado. Buried deep within the earth, these two units deliver 60 MW of electric power each. The large cylinders enclose the electric generators. Below the floor surface are the hydraulic turbines connected to the generators by large shafts. The smaller cylinders on top of the generators are small *exciter* generators which provide power to the electromagnets discussed in Section 2–3. Morrow Point Dam, for which an exterior view is shown in Figure 4–2, cost $60 million to build. (*Courtesy Bureau of Reclamation, U.S. Department of the Interior.*)

We can classify hydroelectric facilities in terms of their capacity to store water. Another way to state the same principle is to give the length of time over which facilities can smooth out variations in stream flow. One class of project is the "run-of-river" project. Essentially, it uses water as it arrives at the facility. Its reservoir usually has limited storage—for perhaps daily or weekly stream flow regulation. An example of this type of project is the Bonneville facility on the lower Columbia River in the Pacific Northwest. It is significant that Bonneville is on the lower part of the river, which benefits greatly from the flow regulation provided by upstream storage dams, including Grand Coulee Dam.

"Storage" projects, a second class of facility, provide water storage for longer periods of time, usually between annual rainfall seasons, although some are built to accommodate more stored water in case of dry years. Examples of the latter are Hungry Horse in Montana, Glen Canyon in Arizona, and some of the projects on the Missouri River. Examples of projects operating on an annual cycle of storage and use are Grand Coulee on the Columbia River, and Fontana in the Tennessee River Basin.

It should not be assumed that the reservoir of a storage project will necessarily hold enough water for continuous operation of all turbines. It is very common that a reservoir will have less water than required for continuous operation; see Problem 3.5. These plants are often used for peaking—that is, for providing power only during peak demand periods, such as during the early evening hours. It is usually quite a challenging task to use just the right amount of water from a reservoir. We wish to use as much hydroelectric power as we can since the "fuel" (water) is free, but we want to save water for peak demand periods. Decisions on when to use water involve expected weather conditions, power demands, fuel prices, size of the mountain snow-pack, and many other factors.

Storage projects can be of two types. In the first type the power plant (with the turbine-generator) is inside, or at the base of, the dam which stores the water. Bonneville Dam, which is an example of this form, has eight large generators rated at 54 MW under a 59 ft. head and two smaller generators rated at 43.2 MW under a

49 ft. head. The turbines are of the Kaplan type with adjustable blades. Hydraulic turbines have different forms, depending on the head and the water flow, to maximize the efficiency. Bonneville's turbines rotate at 75 RPM and discharge water at a rate of 13,300 cfs.

In the second type of storage project, the storage reservoir and the power plant are separate. Water is delivered to the power plant through tunnels or pipes, called penstocks, from the reservoir. The Big Creek #1 facility of the Southern California Edison Company, in the Sierra Nevada, is an example of this type of project. The essential elements of this project, which was built in 1912, are sketched in Figure 3–5. A suitable valley or "bowl" was located in the High Sierras. Three dams were built to hold water flow, forming Huntington Lake, which stores a maximum of 89,166 acre-ft. (an acre-ft. is an acre of water one foot deep). A tunnel and penstocks were built to deliver water stored in Huntington Lake to Powerhouse #1, located near the 5,000 foot elevation. This powerhouse uses a Pelton turbine. This type of turbine, called an impulse turbine, is often used in plants with high heads. The outlet of the

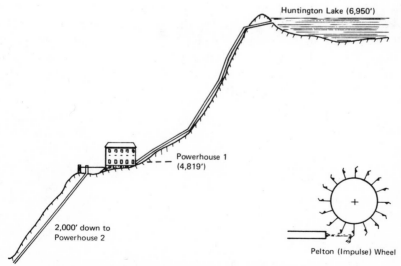

FIGURE 3–5. A High-Head Hydroelectric Project: Big Creek Number 1.

penstock is essentially a nozzle which directs a stream of water at buckets attached to the wheel. The buckets are shaped to make maximum use of the power in the water. When the water falls from the bucket it has transferred almost all of its kinetic energy to the rotating wheel through the force exerted on the bucket. The total power generated by the Big Creek Powerhouse #1 is approximately 81,000 KW.

The water flowing out of Powerhouse #1 is stored briefly in a small reservoir before flowing down a second tunnel–penstock complex to Powerhouse #2, about 2,000 feet lower in elevation. This same water is used four more times to generate electric power before it approaches sea level. The cumulative head is close to 6,200 feet. The result is that more energy per cubic foot of water is generated at Southern California Edison's Big Creek project than at any other hydroelectric project in the world. The company calls this "the hardest working water in the world."

Example 3-4

Suppose that Big Creek Powerhouse #1 were run on a peaking basis for six hours a day during all of July. There are two more lakes above Huntington whose water is used to keep Huntington Lake's level nearly constant during the vacation season, but for the sake of this example do not consider them. At the rate of use given above, what percentage of Huntington Lake's storage would be used? Assume a total efficiency e = 80%. Solve Equation 3-5 for Q. One acre-ft. equals 43,560 cu. ft.

$$Q = \frac{11.8\,P_g}{eH}$$

$$= \frac{11.8 \times 81,000}{0.8 \times 2,131}$$

$$= 561 \text{ cfs.}$$

The total volume of water released is:

$$
\begin{aligned}
V &= Q \times \text{Time} \\
&= 561 \times 3,600 \times 6 \times 31 \\
&= 375,000,000 \text{ cu. ft.} \\
&= 375,000,000/43,560 = 8,610 \text{ acre-ft.}
\end{aligned}
$$

The percentage of Huntington Lake's storage which is used is:

8,610/89,166 = 9.67%

3.5 PUMPED-STORAGE PROJECTS

We come now to a very interesting variation of the hydroelectric project, the pumped-storage project. It is a hydroelectric facility to the extent that it uses water stored in reservoirs to generate electricity with conventional hydraulic turbines. The variation comes from the fact that it does not use the rain cycle to develop the water head. Instead it uses energy generated by some other source to pump the water "uphill" into a reservoir where it can be stored for later use. The basic scheme is illustrated in Figure 3-6.

Recall that the demand for electric energy changes throughout the day (Figure 1-3). Some pumped-storage projects are used to smooth these daily variations (a few have been designed to smooth longer variations). The thermal plant (atomic or fossil-fueled) in the upper left corner of Figure 3-6 supplies power to the city at a nearly constant rate. It is generally not economically desirable to shut down such plants partially or totally for short periods each

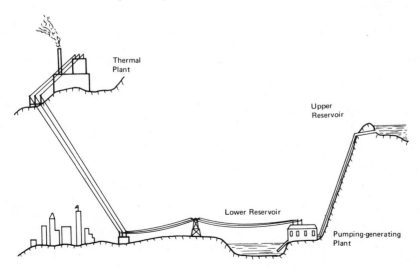

FIGURE 3-6. Pumped-Storage System.

day. During the late night hours the power needs of any area decrease significantly, and the thermal plant has excess power available which is used to pump water uphill from the lower reservoir to the upper reservoir of the pumped–storage facility. On the following day, at the time of peak customer demand, water in the upper reservoir is released to generate the power demand which exceeds the capacity of the thermal plant. Usually the same unit is used to pump water uphill and to generate electricity when the water falls back downhill.

The efficiency of a pumped–storage plant is the product of the efficiencies of the pumping stage and the generating stage. As a rule of thumb, a pumped–storage facility requires 3 KWH of pumping energy for every 2 KWH of generated energy (equivalent to an overall efficiency of 67 per cent). While these losses may be acceptable economically, it is possible that the resulting negative environmental effects will be more than can be accepted. These points will be discussed later.

A number of pumped–storage projects exist now in the United States, including, for example, Taum Sauk (400 MW) in Missouri, Northfield Mountain (1,000 MW) in Massachusetts, and Kittatinny Mountain (330 MW), adjacent to the Delaware River.

√3.6 MULTI–PURPOSE HYDROELECTRIC PROJECTS

Many electric energy generating plants are a part of multi–use or multi–purpose projects. There is increasing interest in multi–purpose developments to justify the use of diminishing resources, and to accomplish more than one objective. Hydroelectric plants have often been involved in such projects, and there is every evidence that this practice will continue.

Objectives which may be combined with a hydroelectric plant include:

1) Flood control
2) Irrigation
3) Recreation
4) Public water supply
5) Navigation

When multi-purpose projects are developed it is necessary that the needs of each area be given adequate attention, and the final design must be an acceptable compromise in light of all of the objectives. This is often a difficult task because needs commonly are in conflict, and it is difficult to put an equitable measure of value on objectives in different fields. As an example, a project may be designed for both pumped-storage power generation and water recreation (swimming, boating and fishing). It might be desirable from the power generation standpoint to have a large daily drawdown of the reservoir. (Drawdown is the decrease in water level as water is released to flow through the turbines.) But the large drawdown would very possibly decrease or eliminate the value of the reservoir as a recreational facility. Hence, a compromise must be effected in the project design to permit a drawdown acceptable to both needs.

For specific examples of multi-purpose hydroelectric projects we refer to some of the examples discussed earlier in this chapter.

A project with a wide range of uses or related activities is the Big Creek facility of the Southern California Edison Company discussed above. See Figure 3-5 for a portion of the project. Besides providing for generation of 690 MW, Big Creek also uses its storage dams and associated reservoirs to aid in flood control and to provide irrigation for the San Joaquin Valley. Huntington Lake and Shaver Lake, which are reservoirs for the project, are major mountain recreation centers. The Company maintains an excellent camping facility (with *electric* stoves) on Shaver Lake, as well as a small fish hatchery to help stock the lakes, and a tree farm.

An additional interesting feature of the Big Creek facility is the use of a ground-based cloud-seeding program intended to increase precipitation. Twelve cloud-seeding generators are located at various ground points in the Sierra Nevada and its foothills (six are quite remote and are fully automatic). The generators burn Propane gas with a Silver Iodide-Acetone mixture producing Silver Iodide particles which rise through moisture-laden clouds. The particles act as water-attracting nuclei. If sufficient moisture is attracted by the nucleus, its weight will eventually cause it to precipitate. Results of the program are difficult to evaluate, but the company feels that the possible very significant increase in stored water

justifies the rather small expense involved. Essentially unanswered
are questions concerning the side effects of the release of Silver
Iodide into the atmosphere, and the loss of potential rainfall for
individuals further along the "cloudpath."

This example raises again the point mentioned earlier about
the difficulty of comparing certain values or objectives. Some costs
are fairly easy to obtain. As we shall see shortly, we can deter-

FIGURE 3–7. San Luis Dam and Power Plant, California State Water Project.
The San Luis facility is part of a huge multipurpose project in California providing public water supply, irrigation, power production, flood control, and
recreation. Water is impounded in relatively wet Northern California and
transported by canal to Southern California. It is used on farms between the
two. At San Luis, over two million acre–feet of water are stored 300 feet
above the power station. The plant, shown in the center of this photograph,
is used for pumped–storage generation. It has a capacity of 424 MW. (*Courtesy
Bureau of Reclamation, U.S. Department of the Interior.*)

mine the cost of electric power generated by a project like Big Creek, and we can compare it with alternative generation schemes. We may be able to obtain a dollar value for irrigation water and perhaps even for flood control. But how shall we measure the value of a swim in a mountain lake, or sunset from a rowboat, or the value of the wilderness we have lost by putting in the project? We have no answers today, but we may have to find answers in the years ahead as the competition for natural resources continues to grow. In the past, power projects have been developed largely in light of criteria which were easily expressed in economic terms. Recreational facilities were often added after the project had met its primary economic objectives. In the future, as man's demand for recreational facilities increases, these needs may come to play a more significant role in the setting of project criteria.

A second example of a multi-purpose project is the Castaic Pumped-Storage Project. This project uses water being pumped from Northern California to Southern California as part of the California Water Plan. (See Figure 3-7.)

Other examples of multi-purpose water projects include the Columbia Basin Project in Washington State, the Tennessee Valley Project, and the Missouri River Basin Project.

Now we turn our attention to the environmental considerations associated with hydroelectric facilities. Rather than speak in generalities, we consider a single major river; we see through its developments some of the effects which hydro projects can have on river systems. Our subject is the Columbia River in the Pacific Northwest.

3.7 THE COLUMBIA RIVER

The headwaters of the Columbia River lie far up in the Canadian wilderness in a valley flanked on the east by the Rocky Mountains and on the west by the Selkirks. The great winter storms that sweep in from the Gulf of Alaska across Washington State and British Columbia provide ample moisture for the snows of winter that become the raging waters of spring.

The river flows north for more than 200 miles before it flanks the barrier presented by the Selkirks. It then turns south and with only a jog or two along the way flows down into the United States and on to the Oregon border, where it suddenly turns west and surges 300 miles to the Pacific Ocean at Astoria.

Before the dam–builders arrived, the Columbia had a powerful flow, sometimes churning, sometimes falling, sometimes placid. It knew great annual salmon runs, and the Indians came to fish at Celilo Falls and Kettle Falls. It had a life that matched the wild country it traversed. But its life, its power, and its flow were the very qualities which made it a river to be harnessed. The Columbia does not fall a great amount in short distances, but it does carry much water to the sea. Its head (H) is small, but its flow rate (Q) is large.

The first major obstacle to the flow of the Columbia was the Rock Island Dam, begun in 1931, near Wenatchee, Washington. Two years later, the Federal Government began construction of Bonneville Dam, the first of a series of major federal dams on the river. Although Bonneville is not one of the more powerful plants, it is important because it represents the emergence of the Federal Government in the building of Columbia River dams and the establishment of the Bonneville Power Administration, or BPA. The BPA has become the major transmitter and controller of power in the Pacific Northwest.

In 1941 the Bureau of Reclamation began construction of what became the largest power plant in the world at the time of its completion, Grand Coulee Dam. Today the combined capacity of the plant is over 2,000 MW. In the late 1960's construction began on a new addition which, if fully authorized, will have a final capacity in 1990 of almost 10,000 MW. This represents about three percent of the total capacity in the United States in 1970.

Between Bonneville Dam, closest to the Pacific, and Grand Coulee, which is the furthest dam upstream which produces power, lie nine other major dams. The period that saw the growth of this system was one of economic optimism, characterized by a belief in the intrinsic value of growth. The spirit of enthusiasm was caught by the great American folk singer and writer Woody Guthrie, who, working in a federally–funded project, wrote 26 songs about the Grand Coulee Dam. Perhaps the best-known of Guthrie's songs on this subject is "Roll On, Columbia," quoted below.

> Green Douglas firs where the water cut through.*
> Down her wild mountains and canyons she flew.
> Canadian Northwest to the ocean so blue,
> Roll on, Columbia, roll on!
>   Roll on, Columbia, roll on.
>   Roll on, Columbia, roll on.
>   Your power is turning our darkness to dawn,
>   So, roll on, Columbia, roll on!
> And on up the river is Grand Coulee Dam,
> The mightiest thing ever built by a man,
> To run the great factories and water the land,
> It's roll on, Columbia, roll on. . . .

The factories were run, the land was watered, floods were tamed, the darkness became dawn. But a price had to be paid. The once-mighty river became a series of long, narrow lakes, formed behind dams. Valleys were flooded and towns lost forever. The salmon runs were cut by nearly 50 percent despite elaborate fish ladders designed to let the fish climb every dam up to the insurmountable Grand Coulee—the end of the line.

For uncounted centuries before the white man came to the Pacific Northwest, Indians had fished the river, particularly at Celilo Falls, where the rocks jutted out into the churning, falling water. The Indians built crude rickety platforms over the water and speared the salmon as they swam upstream. It was one of the most extraordinary and unique features of the area. It belonged to the Indians, and their rights were later ensured by treaty.† As the waters slowly rose behind the Dalles Dam, the second dam up-

---

*ROLL ON, COLUMBIA. Words by Woody Guthrie. Music based on "Goodnight Irene" by Huddie Ledbetter and Alan Lomax. TRO—© Copyright 1957 & 1963 LUDLOW MUSIC, INC., New York, N. Y. Used by permission.

†The reader may wish to read a fictional account of the negotiations for Celilo Falls between white and red men in Ken Kesey's ONE FLEW OVER THE CUCKOO'S NEST.

stream from the ocean, Celilo Falls was lost to Americans, red and white alike. Today, by the edge of a quiet lake where Celilo once was, stands a small brass commemorative plaque.

3.8 DISADVANTAGES OF HYDROELECTRIC POWER GENERATION

Hydroelectric power is usually inexpensive and can often be combined with other desirable objectives, as outlined in Section 3.6. But its costs to the environment are not small, and they will no doubt demand accounting in our decisions for the future. In brief summary, some of the major disadvantages to the environment are listed below.

1) Hydroelectric reservoirs flood regions which may have more desirable uses.
2) Fish migration is restricted.
3) Fish health is affected by changes in water temperatures, and by inadvertent insertion of excess Nitrogen into the water at spillways.(1)
4) Available water and water temperatures may be affected by evaporation from reservoirs and water–release practices.
5) Reservoirs alter stream silt–flow patterns and may eventually fill up with silt themselves and thereby become useless as reservoirs.

3.9 CONCLUSIONS

Many more hydroelectric power plants will be built in the years ahead. There are many potential sites.(2) The rate of growth, however, will be much less than that of thermal plants. The result is that hydro will represent an ever–smaller percentage of total capacity as time passes. The slowing of hydro growth rate is primarily caused by the decreasing number of acceptable sites and

FIGURE 3—8. Grand Coulee Dam on the Columbia River. Grand Coulee Dam, "The mightiest thing ever built by a man," was once the largest electric–power generating plant in the world, turning out over 2,000 MW. Grand Coulee has since been eclipsed in size many times. In 1967, engineers blasted away the right 200 feet of the dam and began construction of a new power plant, shown on the left. Initial units should go into operation in 1975. If plans are completely authorized by Congress, they will lead to a plant with a total capacity of nearly 10,000 MW by 1990. (*Courtesy Bureau of Reclamation, U.S. Department of the Interior.*)

also to increasing environmental pressures against the building of new dams.

We have not yet discussed the economics of hydroelectric power. Before we consider some particular examples, we turn in the next chapter to some basic ideas concerning the costs of power plants, houses, and other things.

REFERENCES FOR CHAPTER 3

1. "Nitrogen Supersaturation Problem," Corps of Engineers, U.S. Army, North Pacific Division, Portland, Oregon, June 24, 1971.
2. "Hydroelectric Power Evaluation," Federal Power Commission, U.S. Government Printing Office, 1968.

ADDITIONAL READING FOR CHAPTER 3

1. D.H. Redinger, *The Story of Big Creek*, Angelos Press, Los Angeles, California, 1949.

   This is the story of the building of the Big Creek Project mentioned in this chapter. It was written by an engineer who helped build the necessary railways and water tunnels over 60 years ago. It is a fascinating historical summary of a great engineering feat which should be interesting to engineer and non-engineer alike.

2. J. McPhee, *Encounters with the Archdruid*, Farrar, Straus and Giroux, New York, 1971.

   This is a trilogy of separate encounters between David Brower, former Director of the Sierra Club and the "Archdruid" of the title, and three "developers." The first and last encounters make the best reading. The last takes place on a raft floating down the Colorado River. The encounter is with an engineer who helped build the huge Glen Canyon Dam on the Colorado. The dialogue is a superb debate over energy needs and wilderness preservation.

3. O. Bullard, *Crisis on the Columbia*, The Touchstone Press, Portland, Oregon, 1968.

   This book discusses the history of the development of the Columbia River, emphasizing the resulting environmental problems. It is highly readable and graphic.

4. J.J. Doland, *Hydro Power Engineering*, Ronald Press, New York, 1954.

   For the reader interested in more complete engineering details on hydro power, this is a good though somewhat dated survey of engineering methods and practices.

PROBLEMS FOR CHAPTER 3

General Problems

3.1. If 0.1″ of rain falls on a square mile of ground, what is the number of cubic feet of water which has fallen? What is the weight of this water? What is the potential energy with respect to the ocean if the ground is at an elevation of 1,000 feet? (232,000 cu. ft.; 14,470,000 lbs.; 14,470,000,000 ft.-lbs.)

3.2. If the efficiency of a hydraulic turbine is 90% and an electric generator connected to it has an efficiency of 95%, what is the overall efficiency of the system?

3.3. The Bureau of Reclamation is presently building a huge new third hydroelectric power plant at Grand Coulee Dam in Washington State. Initial Congressional authorization permits six turbine–generator units, with the first three scheduled for completion by late 1974. Each turbine will have a nameplate rating of 820,000 horsepower, with a 285-foot head. The penstocks will have a diameter of 40 feet; they will discharge 30,000 cubic feet per second to produce the power specified. What is the turbine efficiency? (85%)

3.4. Consider the data of Problem 3.3 above. Assuming a generator efficiency of 98%, what is the electric power generated by each unit? (600,000 KW)

3.5. Consider the data of Problem 3.3 above. Lake Roosevelt, the reservoir behind Grand Coulee Dam, has a total capacity of 9,562,000 acre-ft. If one draws down 20% of this capacity, how long can he run a single unit as described? Why will the answer be approximate? (32 days)

3.6. Consider the data of Problem 3.5 above. If all six 600,000 KW units work at once, and one permits a draw down of Lake Roosevelt of 20%, how long can he run the units? What does this imply about the use of the plant for base or peak loads?

Advanced Mathematical Problems

3.7. Assume that you wish to store a certain fixed amount of gravitational potential energy. Clearly, different combinations of W and H can be used, with the condition that the product must equal the desired value of P.E. Plot a curve of W vs. H for a fixed P.E. of 100 ft.-lbs. Repeat for P.E. equal to 200, 300, 400, and 500. The result is a "family" of curves, each of which is identified by a "parameter," its P.E. What is the shape of these curves?

3.8. Consider the problems stated in 3.7. Assume that the cost of building the reservoir is $C = 2H^{3/2}$. Write an equation for the cost as a function of P.E. and W. If W must lie between 20 and 50 feet, what is the minimum possible cost of the project?

3.9. For the system of Problem 3.7, assume the cost is $C = 2H^{3/2} + W$. If P.E. is to be 200 ft.-lbs., what is the minimum possible cost of the project, and what are the corresponding values of H, W, and C? (H = 5.36; W = 37.3; C = 62.1)

Advanced Study Problems

3.10. Read Part 3 of *Encounters with the Archdruid* (see "Additional Readings" above). Indicate the strongest arguments of Brower and Dominy. Show how these arguments relate to particular needs or desires of people. In your opinion, who wins the "debate"?

3.11. Identify a fairly complex multi–purpose water project. Indicate who benefits from the project, and how they benefit in your region. Attempt to place a quantitative (perhaps a dollar–) value on the good accruing to each person or group.

3.12. Read a history of the Tennessee Valley Authority (TVA), or the Bonneville Power Administration (BPA). Indicate why the Federal Government entered into the projects, and what advantages and disadvantages have resulted.

4

THE COST OF GENERATING ELECTRIC ENERGY

If there were dreams to sell,
Merry and sad to tell,
And the crier rung the bell,
What would you buy?

Thomas Beddoes

The purpose of the economic study of any project is to determine if the project *ought* to be carried out. At one time this study was restricted mostly to factors relating to income and expenses in dollars. Today, however, there is growing concern for the consideration of less tangible factors, such as the effect on the environment. It is necessary that we find a way to include these factors, although not necessarily in precise dollar values.

A recent government document breaks the economic justification of a project into two parts:(1)

1) Economic analysis
2) Financial feasibility

4.1 ECONOMIC ANALYSIS

The economic analysis compares all of the benefits of a project with its costs. The benefit–to–cost ratio must exceed one. Financial feasibility is determined by comparing the income or revenues in dollars with the estimated costs in dollars (usually on an annual basis). Some projects meet one criterion but not the other. In some cases a project which is economically justified but not financially feasible may actually be carried out. As an example, the United States space program was judged to be economically justified, for complex reasons including scientific discovery, national prestige, military development, and others. It is not, however, financially feasible, as the projected "income" is negligible. On the other hand, damming Yosemite Valley for a hydroelectric project might well be financially feasible, but we hope it would never be considered economically justified because of the value of the magnificent scenery. We shall now consider these two parts of the justification study in detail.

As the government report mentioned above suggests, "A project is properly formulated and economically justified if:

1) Project benefits exceed project costs
2) Each separable segment or purpose provides benefits at least equal to its costs
3) The development provides maximum net benefits compared to alternatives
4) There is no more economical means of accomplishing the same purpose which would be precluded from development if the project were undertaken."

The first of the criteria above requires that *all* benefits exceed *all* costs. The difficulty, of course, is that it is extremely hard to assess the value of some benefits and costs, such as a lake to swim in, a stream to fish in, homesites, local weather effects, and many more. For the time being, perhaps, we shall have to be content to consider such benefits and costs in our evaluation, without attempting to ascribe dollar values to them. It may be that in the future we can quantify these values; it may also very well be that man will choose not to attempt to put numbers on such values. Perhaps it would be a mistake to try to quantify them. This problem will not be resolved easily. There is, of course, increasing interest in these intangibles. In large measure we consider such factors in a political framework with final project decisions being made or accepted by a state legislature or commission, or Congress, or the President, or a body such as the Federal Power Commission.*

The second criterion above requires that each separable part of the project be justified. Many parts of a project are not separable. The building of a storage reservoir and the formation of a fishing lake cannot be separated. However, the building of a fish hatchery on the lake could probably be separated. This criterion simply says that if parts *can* be separated, they should be justified independently. This makes it possible to develop justifiable parts of a project and omit parts which cannot be justified.

The third criterion concerns scale. The scale or size of the project should be such that the net benefits are as large as possible. Scale is critical to the success of a project. A hydroelectric facility which is very small for a given site will be uneconomical since little power can be generated, and certain initial and fixed costs will exist even though the cost of the plant may be low. On the other hand, too large a plant will be very expensive and unable to work close to its capacity with a limited supply of water.

An interesting example of trade-offs related to scale arose during the construction of the San Onofre Nuclear Power Generating Plant between San Diego and Los Angeles, California. The site of the plant is a bluff about 100 ft. high overlooking the Pacific Ocean. It was decided that part of the bluff would be excavated to permit the plant to be built close to the ocean. One reason for doing this was to reduce the pumping head required for pumping ocean water to the plant's cooling condensers. The power required to pump a fixed rate of water is proportional to the height it is pumped. This power costs the power plant money (because it cannot be sold), so it is quite desirable to minimize the pumping head. However, decreasing the head means increasing the amount of ex-

*The existence of final decision-making bodies does not free either citizens or engineers from pointing out and evaluating all possible benefits and costs, both tangible and intangible.

cavation and also increasing the possibility of damage from the ocean. In the case of the San Onofre plant it was determined that an optimum plant elevation was about 20 feet above the ocean.

Example 4-1

(This example is based on the type of problem mentioned in the preceding paragraph. It does not, however, refer directly to the San Onofre situation nor does it use numbers taken from that project.) Assume that the cost per 1,000 KWH for pumping cooling water to a plant with a pumping head of H is:

$$C_p = 0.02 \text{ H dollars}/1,000 \text{ KWH}$$

and the cost of excavation, including an estimate of the probability of ocean water damage, is:

$$C_e = \frac{1,600}{H^2} - 0.16 \text{ dollars}/1,000 \text{ KWH}$$

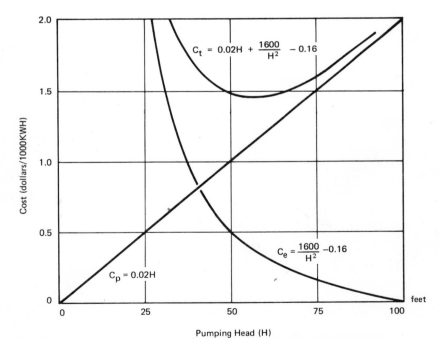

FIGURE 4–1. An Optimum Scale Study.

These curves are plotted in Figure 4-1, along with their sum C_T.

$$C_T = 0.02H + \frac{1,600}{H^2} - 0.16$$

We wish to make this total cost a minimum. This can be done graphically by observing the minimum point of C_T. C_T is a minimum about where H = 54 feet. (For those who have studied calculus this result can also be obtained by setting the first derivative of C_T to zero and solving for H.)

The last of our four criteria requires that there not be any more economical means of accomplishing the same purpose. That is, it must be demonstrated that the project supplies the type of load required better than any of the alternatives, such as a fossil–fuel plant, an atomic plant, or perhaps internal combustion or gas turbine units.

4.2 FINANCIAL FEASIBILITY

We move now from the problem of economic evaluation to a determination of financial feasibility. Here we must compare the dollar cost of producing electric energy with the expected income from the power. The cost of electric energy from any type of generation is usually expressed as the sum of two costs, namely *fixed charges* and *operation and maintenance charges.*

Fixed charges include taxes and insurance, amortization or depreciation, and interest or cost of money. These charges relate to the presence or existence of the plant, regardless of how much power is being produced. They do not vary with production and are hence said to be fixed. We shall discuss them in detail shortly. Operation and maintenance charges include fuel costs (if any), supplies, expenses, supervision, maintenance engineering and labor.

An excellent analogy can be made between the costs of an electric plant and the cost of a home. The latter has taxes and insurance. Amortization and interest charges are accounted for in paying off the mortgage obtained from the lender (usually a bank). These fixed costs do not depend on use of the home. However, variable charges for heating, electricity, upkeep (repairs, painting, etc.) do depend at least partially on the use of the home.

One of the most important costs for an electric power plant is the original cost of acquiring property and constructing a facility. The scale of power plants is such that this requires a relatively large amount of capital investment. It is necessary to have this money available at the beginning of the life of the facility to pay for its construction. The necessary capital is usually borrowed from an investor. It is necessary then to repay the money borrowed plus interest charged for use of the money. This repayment or amortization period is usually 50 years for a hydroelectric facility. Such facilities typically have an expected lifetime near 100 years. However, Federal Power Commission licenses extend only 50 years or less. At the end of this time the company must apply for a new license.

A distinction should be made here between amortization and depreciation. Assume that we borrow money to build some facility (of any type) with an expected lifetime of 50 years. Each year we pay an interest charge for using the money plus an amount necessary to pay off part of the capital. At the end of 50 years we have paid off the loan, and theoretically our plant's value has decreased to zero (due, perhaps, to deterioration, obsolescence, or other factors). Since the plant's value has decreased to zero, we design a new plant, borrow some more money and start over. Suppose, however, that at the beginning of the life of the plant we had sufficient cash on hand to pay for all construction costs; that is, we owned the capital. We might argue that we could therefore forget about the cost of money and cut our expenses sharply for the 50 year life of the plant. But that would not be good economics, because at the end of the 50 years, we would have lost both our plant *and* our capital. We could have loaned out our capital and borrowed some more capital as before. At the end of the 50 years we would have our capital back plus interest. Of course we might also find a way to lend ourselves the money and pay ourselves back. One way to do this is to pay for the plant ourselves and then pay a fixed amount each year into a fund established to counteract the depreciation of the plant. This can be

FIGURE 4–2. Morrow Point Dam, on the Gunnison River in Colorado. The primary factor in the cost of hydroelectric power is the cost of the dam and powerhouse. The Morrow Point Dam cost $60 million. The spectacular 360 foot waterfall illustrates another economic consideration. This water is wasted as far as electric energy generation, because it does not pass through the turbines. More turbines could have been installed to handle the overflow during those short periods when the dam cannot contain all the water of the Gunnison. But more turbines cost more money. So we must seek an economic compromise between turbine cost and the value of the additional energy which could be generated. (*Courtesy Bureau of Reclamation, U.S. Department of the Interior.*)

called a depreciation fund, or sinking fund. It is a fund into which payments are "sunk" to produce a desired amount at the end of a set period.

It should be clear from the above that a project must pay for the capital investment, one way or another, whether we borrow the money from an outsider or from ourselves. If it does not, the payments are not a true reflection of the cost. Sometimes people might wish to pay into both amortization *and* depreciation funds. This would be an abuse in the other extreme. In this case the ratepayer would be paying not only for the present plant but also for its replacement, simultaneously. This would be an unfair burden.

The Soviet Union once attempted to ignore the problem of capital cost by assuming capital belonged to the state and need not be accounted. The result was an incorrect decision on development of a project, and later the decision had to be reversed. This case is discussed in an interesting short book on engineering economics by Sporn.(2)

What is interest? Interest is money we must pay for the privilege of borrowing money. It is usually expressed as a percentage rate per year. An interest rate of 5 percent means that we must pay $5 for every $100 we borrow for each year we have the "capital." People lend money because they have it and are willing to forego the use of it temporarily in order to acquire more money (interest), and set aside "savings" for some future need. People borrow money and pay interest because they expect to produce goods (such as tennis rackets or electric power) which return an income greater than the cost of the investment. If they fail to do so they will go out of business.

How much does it cost to borrow money? Suppose you wish to borrow $100,000,000 (a fairly reasonable price) to build a 300 MW hydroelectric plant, paying off the loan over 50 years. How much should you pay each year? The answer depends on the interest rate and the method of repayment chosen. There are a number of methods of repaying a loan. The most common is to make equal payments at the end of each of n years so that the loan is completely paid off after n years. What should these equal payments be? Let's call the original loan value P. Later we will let P equal $100,000,000 to answer the question above. We have borrowed P dollars at an interest rate i. At the end of the first year the value of the money has become:

$$P(1 + i) \qquad\qquad (4\text{-}1)$$

Now we make a payment toward paying off the loan. If we decide to pay it off in one year we simply let our payment (call it R) be $P(1 + i)$. Suppose, however, we choose to pay off our loan in two years. Then we make a payment R, which, when subtracted from the value of the money at the end of one year, leaves a loan balance:

$$P(1 + i) - R$$

At the end of the second year we pay i interest on this balance. Again we subtract the same R, and the result must be zero to pay the loan off in two years.

$$[P(1 + i) - R] (1 + i) - R = 0$$
$$P(1 + i)^2 - R(1 + i) - R = 0$$

Similarly, the equation for a three year repayment plan yields:

$$P(1 + i)^3 - R(1 + i)^2 - R(1 + i) - R = 0$$

and for an n–year repayment plan:

$$P(1 + i)^n - R[1 + (1 + i) + \ldots + (1 + i)^{n-1}] = 0 \qquad (4\text{-}2)$$

If we can evaluate the sum in the brackets we can find R, our annual payment, for any values of P and n. Write the sum in the brackets as:

$$S = 1 + a + a^2 + \ldots + a^{n-1}$$

where $a = 1 + i$.

The series $(1, a, a^2, \ldots)$ is very common; it is called the geometric series. Its sum is easily found. Multiply S by a, subtract the two sums, and solve for S.

$$S = 1 + a + a^2 + \ldots + a^{n-1}$$
$$aS = \quad\; a + a^2 + \ldots + a^{n-1} + a^n$$

$$\overline{S(1 - a) = 1 - a^n}$$

$$S = \frac{1 - a^n}{1 - a} \tag{4-3}$$

We can now use this result to help us solve for R in equation (4-2).

$$R = \frac{P(1 + i)^n}{[1 - (1 + i)^n]/[1 - (1 + i)]}$$

$$= \frac{Pi(1 + i)^n}{(1 + i)^n - 1} \tag{4-4}$$

Example 4-2

If we borrow \$100,000,000 for our 300 MW hydroelectric facility, what must our yearly payments be if the interest rate is 7% and the repayment period is 50 years?

This problem requires a direct application of (4-4). It can be solved with some types of slide rule, with logarithms, or through use of a table. We will not want to develop a table for specific values of P, since the multiplication step is quite simple. Hence we form a table of the multiplier of P. Call it r.

$$r = \frac{i(1 + i)^n}{(1 + i)^n - 1} \tag{4-5}$$

This function is given in Table 4-1 for various values of i and n. Entering Table 4-1 for i = 7 percent and n = 50 years, we find r = 0.072. Hence our yearly payment for interest and amortization is:

$$R = 100,000,000 \times 0.072 = \$7,200,000$$

(The reader should realize that the three-place accuracy of Table 4-1 is not adequate for calculations of payments in most actual situations, since the errors involved may amount to thousands of dollars if P is large enough. More complete and accurate tables are to be found in texts on engineering economics(3) as well as other sources.)

TABLE 4-1.

Values for the Fraction $r = \dfrac{i(1+i)^n}{(1+i)^n - 1}$

n \ i	1	2	4	5	6	7	8	10
1	1.010	1.020	1.040	1.050	1.060	1.070	1.080	1.100
2	0.507	0.515	0.530	0.538	0.545	0.553	0.561	0.576
3	0.340	0.347	0.360	0.367	0.374	0.381	0.388	0.402
5	0.206	0.212	0.225	0.231	0.237	0.244	0.251	0.264
10	0.106	0.111	0.123	0.130	0.136	0.142	0.149	0.163
15	0.072	0.078	0.090	0.086	0.103	0.110	0.117	0.132
25	0.045	0.051	0.064	0.071	0.078.	0.086	0.094	0.110
30	0.039	0.045	0.058	0.065	0.073	0.081	0.089	0.106
40	0.030	0.037	0.051	0.058	0.067	0.075	0.084	0.102
50	0.026	0.032	0.047	0.055	0.063	0.072	0.082	0.101
75	0.019	0.026	0.042	0.051	0.061	0.070	0.080	0.100
100	0.016	0.023	0.041	0.050	0.060	0.070	0.080	0.100

Example 4-3

If you take out a $40,000 mortgage on a house, agreeing to pay 6 percent interest, and repay the loan in 25 years, what is your annual payment? What is the corresponding average monthly payment? What is the total payment over 25 years?

From Table 4-1, the repayment multiplier r is 0.078. Hence the yearly payment is

$$R = 0.078 \times \$40,000 = \$3,120$$

This corresponds to an average monthly payment of $260. Over 25 years you pay $78,000.

This completes our study of fixed costs. We have noted the existence of taxes and insurance as costs, and we have studied one method for calculating the cost of capital investment. We turn now to the question of operation and maintenance charges.

Fuel costs will be the most important operation charge in a fuel-burning plant. In the case of hydroelectric plants, however, fuel costs are neglected. These plants typically have high initial investment costs, and hence high fixed charges, but relatively low operation and maintenance charges. Maintenance is lower for hydroelectric plants than for fuel-burning plants because of the relative simplicity and reliability of the former.

Once all of the costs, fixed and non-fixed, of the proposed plant are identified, we can proceed to find the net cost of producing electric energy. We usually express our final answer in mills/KWH. If this figure is equal to or less than the rates we presently charge or anticipate charging, the project is financially feasible.

We conclude this section with a typical example. The figures given are not taken from a specific case but are nonetheless representative of projects of this size.

Example 4-3

We wish to build a 200 MW hydroelectric plant for a total project cost of $71,000,000. The interest rate is 6 percent, and the amortization time is to be 50 years. Taxes and insurance are estimated to be $1,000,000 per year. Operation and maintenance costs will be about $1.60 per kilowatt. It is assumed that the facility will have a plant factor of 40 percent, which is the ratio of the average load on the plant to the power rating of the plant. Determine the cost of producing energy.

A quick calculation shows that the plant cost per kilowatt is:

$$\frac{71,000,000}{200,000} = \$355$$

Hydroelectric plants typically have plant costs of $250 to $500 per kilowatt. This is high as electric plants go, but is compensated by the lower operating costs.

Now we determine the number of KWH produced per year. There are 8760 (24 × 365) hours in a year. The number of KWH per year is the power rating of the plant times the plant factor times 8760.

Energy = 200,000 × 0.4 × 8760 = 700,000,000 KWH

We find next the cost of borrowing $71,000,000 at 6 percent for 50 years. From Table 4-1, the equal payment rate is seen to be 6.3 percent of the original capital.

0.063 × 71,000,000 = $4,473,000

Total annual fixed costs (cost of money plus taxes and insurance) are

$$\$4{,}473{,}000 + 1{,}000{,}000 = \$5{,}473{,}000$$

and the resulting cost per KWH is:

$$\frac{5{,}473{,}000}{700{,}000{,}000} = 7.82 \text{ mills/KWH}$$

Operation and maintenance costs are:

$$\$1.8 \times 200{,}000 = \$360{,}000/\text{year}$$

and the corresponding energy cost is:

$$\frac{360{,}000}{700{,}000{,}000} = 0.514 \text{ mills/KWH}$$

Hence, total energy costs are:

Fixed Costs	$7.82
Operation and Maintenance	0.51
	8.33 mills/KWH

It should be apparent that a higher plant factor would reduce the energy costs. The usual reason for the relatively low plant factor in a hydroelectric plant is the lack of water. In a particularly wet year this factor may rise significantly and cut the cost of producing power. As we shall see in the next chapter, the plant factor is a little less important in reducing the cost of fossil–fuel generated energy because the more energy you produce, the more fuel you need to produce it and the more the fuel costs.

4.3 FIXED AND VARIABLE COSTS

In this section we study in more detail the role of fixed and variable costs by considering three very different examples. The first is the hydroelectric plant observed in the previous example. The second concerns the cost of heating a house. The third example treats a small household appliance.

In Example 4-4 we saw how three fixed costs contributed to final energy cost. The cost of energy for that case can be expressed generally as:

$$C_E = \frac{R + T + O}{8760fP} \text{ (mills/KWH)} \qquad (4-6)$$

where R is the annual payment on capital and interest, T stands for taxes and insurance, and O for operation and maintenance, f is the plant factor, and P the power capacity of the plant in KW. For the plant in Example 4-4 the energy cost becomes

$$C_E = \frac{3.33}{f} \text{ (mills/KWH)} \qquad (4-7)$$

The cost of energy is inversely proportional to the plant factor. Thus the more the plant is used, the less the cost of energy. Equation 4-7 is plotted in Figure 4-3. In this example all costs are assumed fixed and the only energy costs depend on plant factor.

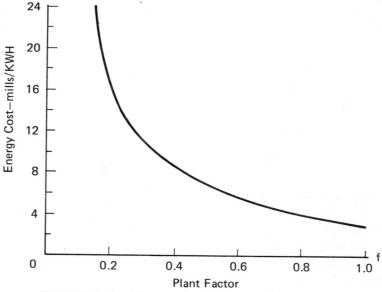

FIGURE 4-3.   Effect of Plant Factor on Energy Cost.

When we study fossil–fuel plants, we add a fuel–cost term which is independent of plant factor. Total energy cost for that situation can be expressed as

$$C_E = \frac{F.C.}{8760fP} + F \text{ (mills/KWH)} \tag{4-8}$$

where F.C. stands for total fixed costs and F represents the cost of fuel. The latter does not depend significantly on plant factor.

For our second example we consider a situation in which capital or fixed costs can be traded off against operating or variable costs.

Example 4-5

In the spring of 1973 the Pacific Gas and Electric Co. of San Francisco announced a new Energy Conservation Home project.(4) According to a PG&E official, such a home ". . . of 1,370 square feet located in San Jose would cost about $140 more to meet Energy Conservation Home requirements, but this home will cost about $20 less to heat each year." Assume that the $140 is added to the mortgage and amortized over 30 years at 7 percent interest. Compare the added cost of the mortgage payment (fixed capital cost) with the savings in operational costs.

From Table 4–1 we see that the repayment factor r is 0.081. Hence the added mortgage payment is:

0.081 × $140 = $11.34 per year

Thus the homeowner pays an additional $11.34 per year or $0.95 per month to the mortgage company, but saves $20 per year or $1.67 per month on the heating bill.

This is an example of energy conservation that saves money without decreasing one's "standard of living." This concept is discussed again in Chapter 16.

Our final example concerns the cost per hour of operating a small appliance.

Example 4–6

Determine the cost per hour of operating a circulating fan. (See Table 1–4.) Assume that the fan costs $20. It will be used for 10 years and then junked, and will require no maintenance (a fair assumption for fans). Assume energy costs $0.02/KWH.

The fixed cost or cost of owning the fan per year is $2. Since the fan requires 88 watts, in one hour it uses 0.088 KWH at a cost of $0.00176. Thus the cost of running the fan per hour is:

$$C = \frac{2}{H} + 0.00176 \text{ (dollars/hour)} \tag{4-9}$$

where H is the number of hours the fan is operated per year.

If the fan is operated 100 hours per year, the cost is $.02176 per hour. If the fan is operated 1,000 hours per year, the cost is $.00376 per hour.

For many devices such as lightbulbs the analysis above would be quite inappropriate, since the life of a lightbulb is highly dependent on the hours of use. This is much less true of a fan.

4.4   THE ECONOMICS OF OTHER POWER PLANTS

In this chapter we have concentrated on hydroelectric power plants, since that is the only type of plant we have discussed as yet. However, the basic principles of this chapter apply to all other kinds of plants, existing and future. The major difference for some plants we shall study (fossil–fuel and nuclear, for example) is that they will have large variable costs for fuels. We shall be extending the ideas of this chapter as we discuss new ways to generate electric energy. The cost factor is extremely important in the analysis of any generation technique. Many proposed schemes which use "free" fuels, such as tidal and solar plants, have failed because of huge capital costs. As new schemes in these areas and other areas are proposed, they must meet the test of financial feasibility as well as overall economic desirability.

FIGURE 4—4. Coal handling facility. This is a 250 MW coal–burning plant in central Missouri. Coal is stockpiled at 1,800 tons per hour by a dual system of 42″ wide belt conveyors. It is delivered to the power plant at a rate of as much as 700 tons per hour. The huge stockpile and the fuel–flow rates indicate the very large quantities of coal which are burned in fossil–fuel power plants. These large quantities of fuel will become a major factor in the economic analysis of steam power plants in the chapters ahead. (*Courtesy of Bulk Handling Systems.*)

REFERENCES FOR CHAPTER 4

1. Federal Power Commission, "Hydroelectric Power Evaluation," U.S. Government Printing Office, 1968.
2. P. Sporn, *Technology, Engineering, and Economics,* MIT Press, 1969.
3. E.L. Grant and W.G. Ireson, *Principles of Engineering Economy,* The Ronald Press Co., New York, Fourth Edition, 1960.
4. *P G and E Progress,* Pacific Gas and Electric Co., San Francisco, Cal., June, 1973.

ADDITIONAL READING FOR CHAPTER 4

1. Reference 2 above is a short and very readable discussion of economic factors in technical development. It requires no background in economics or mathematics.
2. R. Dorfman and N. Dorfman, Eds., *Economics of the Environment*, W.W. Norton & Co., New York, 1972.

   This book contains selected readings and extensive bibliography on a number of general problems in economics related to the environment.

PROBLEMS FOR CHAPTER 4

General Problems

4.1. Explain why there are undesirable qualities in the maximum and minimum sizes or scales of the following examples:

   a) Capacity of a theatre
   b) Loudness of a radio
   c) Size of a corn flake box
   d) Strength of a cable holding up an elevator
   e) Width of a fixed–weight triangular cross–section gravity dam
   f) The Statue of Liberty

4.2. Suppose you wish to buy a house for $35,000. You pay $5,000 down and borrow the rest from the bank, agreeing to make equal annual payments over 30 years. Find your annual payment for interest rates of 4 percent and 8 percent. Find the total amount of money you will pay back to the bank in 30 years for these two rates. ($1,740, $2,670; $52,200, $86,100)

4.3. Consider Example 4–3. What is the cost of energy production in a wet year in which the plant factor is 60 percent? (5.55 mills/KWH)

4.4. A 500 MW hydroelectric plant will cost $150,000,000. Interest rates are 7 percent and the repayment time is 50 years. Taxes and insurance will be $3,000,000/year. Operation and maintenance costs will be $1.25/KW. Plant factor is 50 percent. Find the cost of energy. (6.58 mills/KWH)

Advanced Mathematical Problems

4.5. Show that capital P invested at a rate i compounded annually has a value after n years equal to:

$$S = P(1 + i)^n$$

(Compounding annually means that once each year the value of the accumulated capital is increased by the percentage interest rate.)

4.6. Using the formula of Problem 4.5, find the value of $1,000 invested at 5 percent interest for: 1 year, 5 years, 30 years, 100 years. ($1,050, $1,276, $4,320, $132,000)

4.7. Using the formula of Problem 4.5, find the value of $1,000 invested for 30 years at: 1 percent, 3 percent, 5 percent, 10 percent. ($1,348, $2,428, $4,320, $22,900)

4.8. *Present worth* of some amount of money S available n years in the future is defined as the amount of money which would have to be invested at the present rate of interest to yield S dollars in n years. Present worth is easily found from Problem 4.5, for annual compounding, to be:

$$P = \frac{S}{(1 + i)^n}$$

Find the present worth of $100,000 payable in ten years if that present worth is invested at 6 percent interest. ($55,900)

4.9. Consider the following example of the application of the present worth principle introduced in Problem 4.8. You are building a power plant which will eventually house two generation units, although only one is required today. The second one can be built at a cost of $10,000,000 today or $15,000,000 in five years. If the present interest rate is 7 percent determine whether it is better to put in the generator now or in five years.

Advanced Study Problems

4.10. Make an extensive analysis of the operating costs and other (indirect) costs of an automobile. *After* you have completed your own analysis, consult "Cost of Operating an Automobile," U.S. Department of Transportation, April 1972. (U.S. Government Printing Office.) What part of total cost is the initial capital cost?

4.11. Contact a local power company, or other industry, which is building a new plant. Determine the capital cost, the method of financing, the amortization period, and the part which the capital cost plays in final product cost.

5

FOSSIL-FUEL ELECTRIC POWER PLANTS

Can a man take fire to his bosom and his garments not be burned?

Proverbs

More than one hundred million years ago, nature began the slow process of creating what we know today as the fossil fuels: coal, oil, and natural gas. It began with vegetation, growing in the light-energy of the sun. The vegetation died, became buried, and over countless millions of years fossilized to become the fuels we have on earth today.

Man has used fossil fuels for at least 10,000 years to produce fire, first for light, warmth, and cooking and later for the first steps toward technology. The development of the practical steam engine about 250 years ago began a new era of using nature's fossil-fuel resources to help do man's work.

In this chapter we explore the ways in which we use fossil fuels in steam power plants to produce electric energy.

A steam power plant burns fuel to turn the shaft of an electric generator. Our first goal is to describe how this task is accomplished. With a simple explanation as a starting point, we proceed to explain certain refinements which can make the system efficient and practical. This leads us to a brief look at the thermodynamic limitations of thermal processes. Then we turn to the process of combustion, with particular attention to troublesome and undesired byproducts. These discussions lead quite naturally to a consideration of the environmental problems relating to fossil-fuel plants. We conclude with a discussion of the economics of such plants.

A point of clarification is in order. This chapter is particularly concerned with fossil-fuel plants. Such plants use the following fuels, with percentage contribution to U.S. electric energy production in 1970 indicated in parentheses:

Coal (46%)
Natural gas (24%)
Oil (12%)

Steam power plants include not only fossil-fuel plants, but also nuclear plants and geothermal plants. In this chapter we discuss first those features of steam power plants common to all of the above, and then concentrate our discussion on fossil-fuel plant considerations. In Chapter 6 we take up the peculiarities of atomic steam power plants, and in Chapter 7 we deal with geothermal steam power plants.

5.1   A BASIC STEAM POWER GENERATION SYSTEM

Two very primitive devices serve as examples of how steam power can be harnessed to do mechanical work. The first device, invented about 2,000 years ago by Hero of Alexandria, is sketched in Figure 5-1a. The fire heats the water to boiling. Additional heat produces steam, which travels up one of the arms into the sphere, which is free to rotate. The sphere has two jets pointing in opposite directions through which the steam is allowed to flow. (A

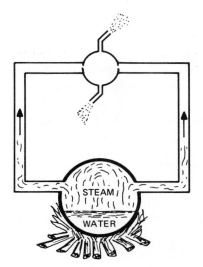

FIGURE 5—1a. Hero's Aeolopile (1st Century A.D.)

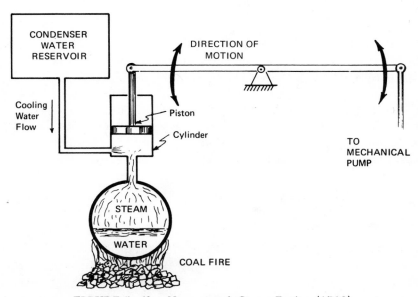

FIGURE 5—1b. Newcomen's Steam Engine (1712).

jet device has a forward thrust or force because some substance is emitted at the back.) The flow of steam causes a reaction, essentially like that of a jet engine, which causes the top sphere to rotate. This is a very simple example of a "reaction" turbine. Hero did not harness this device for any practical use.

Our second example, sketched in Figure 5-1b, is the basis of the first practical steam engine, built by Thomas Newcomen in 1712 to pump water from English coal mines. Coal was burned to produce steam which filled the cylinder above the boiler, forcing the piston to the top. The steam supply was then cut off, and water was sprayed into the cylinder, causing the steam to condense. When the steam condensed, atmospheric pressure outside the cylinder forced the piston down, resulting in the mechanical work used to operate the water pump. (We shall see this use of a spray of water to condense steam in a modern application in Chapter 7 on geothermal power.)

From this modest start, the steam engine has evolved over the past 250 years into a far more complex and efficient device. We trace a part of this evolution to indicate where we have come in the years since Newcomen.(1) Let us start with the simple system sketched in Figure 5-2.

5.2 TOWARD AN EFFICIENT PLANT

The system shown in Figure 5-2 will produce mechanical power which can be used to turn an electric generator. However, it will be extremely inefficient. The efficiency would no doubt be less than one percent, whereas modern steam engines have efficiencies of nearly 40 percent. How then do we alter this simple system, without changing its basic principle of operation, to achieve such efficiencies? We answer this question by considering the system in four parts:

1) The heating subsystem, including the fire
2) The steam subsystem, including the boiler and steam delivery system
3) The steam turbine
4) The used–steam disposal subsystem, including the condenser

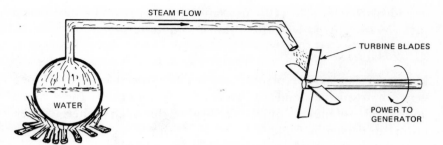

FIGURE 5—2. A simple steam power system.

The heating subsystem consists of those devices associated with getting the fuel to the boiler, burning it efficiently, and removing the products of combustion. The heating subsystem for a fossil-fuel plant is different from those of atomic or geothermal plants, but the other three subsystems are usually very similar if not identical for fossil-fuel, nuclear, and geothermal plants.

Figure 5-2 shows a very simple stationary fire. A practical large-scale fossil-fuel plant requires huge amounts of heat, and a way must be found to move in fuel and air efficiently and to remove combustion by-products. To see the magnitude of this problem, consider how much energy is available in typical fuels, and hence at what rate we must use such fuels.

Table 5-1 gives the "heating value" and the associated electric energy generation value of certain fuels. Heating value is the maximum amount of energy released when fuel combines with oxygen in a combustion process. It is commonly expressed in British Thermal Units (BTU) per pound (coal, oil) or per cubic foot (gas). One BTU is the amount of heat needed to raise the temperature of one pound of water by one degree Fahrenheit.

TABLE 5-1.

Average Heating Values of Common Fuels

Fuel	Heating Value	Electric Energy Generation Rate	Assumed Efficiency
Oil	18,500 BTU/lb.	2.15 KWH/lb.	40%
Coal	12,000 BTU/lb.	1.4 KWH/lb.	40%
Natural Gas	1,000 BTU/cu. ft.	0.12 KWH/cu. ft.	40%
Uranium	200,000,000 BTU/lb.	17,500 KWH/lb.	30%

An ideal heat engine (100 percent efficient) would convert heat into electric energy with the following equivalence:

$$3{,}413 \text{ BTU} = 1 \text{ KWH} \tag{5-1}$$

Modern steam power plants have actual efficiencies near 40 percent for fossil fuels and 30 percent for atomic fuels. These values and the above energy equivalence lead to the electric energy generation rates given in Table 5-1.

Example 5-1

A large modern coal-fired steam power plant, shown in Figure 5-5, was recently built near Centralia, Washington. The plant has a generating capacity of 1,400,000 kilowatts. The heating value of the coal used at this plant is 8,100 BTU/lb. Assuming an efficiency of 40%, find the tons of coal per minute needed to run this plant.

The energy produced each minute is:

$$1{,}400{,}000 \times \frac{1}{60} = 23{,}300 \text{ KWH}$$

The electric energy generation rate for this coal is

$$\frac{8{,}100 \times 0.4}{3{,}413} = 0.95 \text{ KWH/lb.}$$

Hence the amount of coal each minute is

$$\frac{23{,}300 \text{ KWH}}{0.95 \text{ KWH/lb.}} = 24{,}500 \text{ lbs.} = 12.25 \text{ Tons}$$

This example clearly illustrates the enormous quantities of fuel required by coal plants. Large amounts of fuel are also used by plants burning other fossil fuels. It is apparent that large fossil-fuel plants must bring in, burn, and dispose of the waste of a great deal of fuel. Coal is usually brought in by train, or by conveyor belt if the plant is a so-called "mine-mouth plant." Oil is piped in or brought in by ship. Natural gas is presently piped in. However, it is possible to liquefy and transport natural gas at cryogenic temperatures by ship. In late spring of 1973, for example, the United States contracted with the Soviet Union to provide the United States with large quantities of liquid gas to be delivered by ship. In all cases the distance from the fuel source to the plant is an important economic consideration because of the cost of transportation. This factor becomes far less important with atomic plants because of their much greater energy concentration (see Table 5-1).

A very simple boiler or steam generator for a fossil-fuel plant is shown in Figure 5-3. The large outer vessel contains the fire. This vessel can be more than 100 feet in height and have a combustion chamber volume exceeding 100,000 cu. ft. The flame or hot gases heat the water which flows through many miles of tubing inside

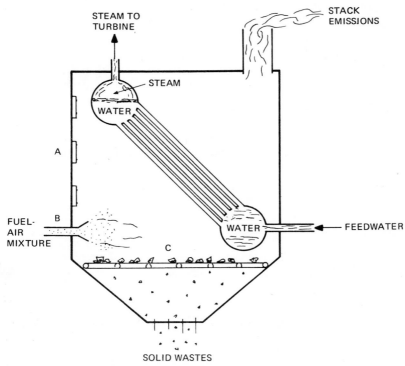

FIGURE 5—3. Boiler (Combustion Chamber).

the chamber. Coal can either be introduced in chunks on a moving grate (C) or it can be blown into the chamber in a pulverized form (B). Oil or natural gas can be blown in through burners (A) on the sides of the walls. The burners look much like burners on a cooking stove, although they are much larger.

Wastes from the combustion process are released as gases from tall stacks and as solid wastes, such as ashes, into railroad cars for dumping.

The heat produced in the combustion process changes the water into steam. The steam flows through large pipes to the steam turbine. After it is used once, it is usually re-heated at least once and used again. We shall return to this process after discussing the steam turbine.

Today's steam engine is a multi-stage turbine. A part of such a turbine is shown in Figure 5-4. Steam enters the turbine on the left. A set of fixed blades (shown without cross-hatching) directs the steam to the moving blades (cross-hatched) which are attached to the main shaft. A typical turbine will have hundreds of blades of different lengths. The size of the turbine chamber increases from left to right—the flow direction of the steam in the drawing— because the steam is expanding as it does its work of driving the blades. In the simple steam engines we have already seen, work and steam expansion always go together. When the steam has done its work it is exhausted from the turbine into the condenser.

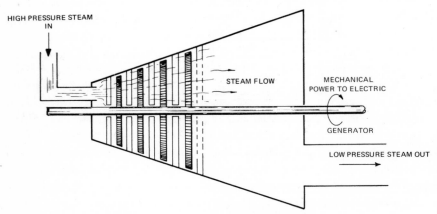

FIGURE 5—4. Modern Steam Turbine.

The purpose of the condenser is to cool the steam, causing it to condense back into water. It is then pumped back to the boiler to be made into steam again. Two very important and interesting questions come up at this point.

1) Why not just let the steam go out into the air and use some new water for new steam, thus avoiding the need to build a condenser?

2) Why cool off the exhaust steam when it is going to be re-heated again by the boiler anyway?

There are at least three reasons for continually re-using the same water for steam. The first is that otherwise huge amounts of steam which would be released to the atmosphere would almost certainly cause very undesirable local weather conditions. The second is that new water would have to be introduced to the boiler continually. But water is often expensive or scarce. Many steam plants are located in areas which would not have new water available. The third reason for re-using water is that the water employed in a steam cycle must be very pure to avoid corrosion and mineral deposits in the system. It would cost entirely too much to have to purify new boiler water constantly. Saving the condensed water means saving money.

The answer to the second question above is related to one of the most fundamental laws of nature which man has been able to conceive: The Second Law of Thermodynamics, which states that it is impossible to construct an engine which, when operating through a complete cycle, will convert all of the heat supplied to it into work.

We use a condenser to cool and condense the steam because the work which can be accomplished by a heated substance operating in a closed cycle depends not only on the temperature of the gas but also on the temperature of the region into which the gas will be exhausted. The greater the difference between these temperatures, the greater will be the work which can be done. We discuss this in more detail in Section 5.3 below.

We shall want to say quite a bit more about the operation of the condenser later, since it is at the heart of one of our major environmental problems. But first we return to the steam cycle.

FIGURE 5—5. Coal–burning power plant at Centralia, Washington. Two 700 MW units make up this huge fossil–fuel steam power plant. This plant, which went on line in 1972, is the first large steam plant in the hydroelectric–rich Pacific Northwest. It burns about 12 tons of coal each minute (see Example 5–1). The plant uses large electrostatic precipitators to eliminate particulate release, and it employs induced-draft cooling towers for condenser–water cooling. (*Courtesy of Pacific Power and Light Co., Portland, Oregon.*)

Recall that steam is generated in the boiler, piped to the turbine, and exhausted into the condenser. The resulting water is pumped back to the boiler. This is a very basic description of the fundamental steps. In fact a number of other things happen to the steam in its cycle in a modern plant. These added features are shown in Figure 5-6, along with most of the basic components already discussed. The primary purpose of most of these features is to increase plant efficiency. The simple system of Figure 5-2 has an efficiency much less than one percent. A modern steam power

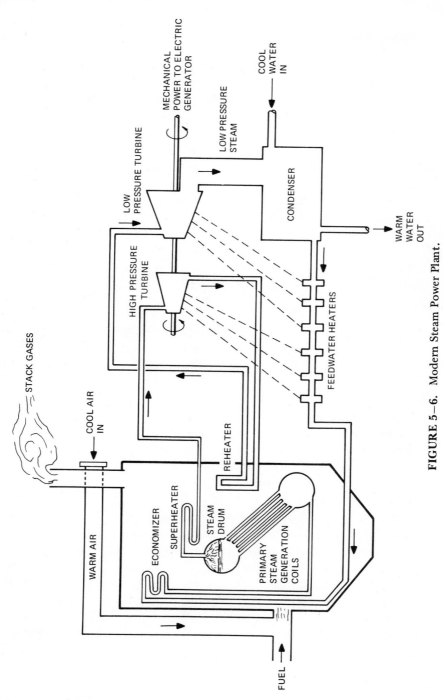

FIGURE 5–6. Modern Steam Power Plant.

101

plant, using the features shown in Figure 5-6, will have an efficiency of nearly 40 percent.

First, consider the combustion system. Combustion requires fuel and air. Air is brought in cold as shown, and it is heated by passing it through pipes in the stack. Heating the air improves the combustion process. We must be careful, however, not to remove too much heat from the stack gases. These gases contain water vapor from the combustion process. If the water condenses, it will combine with Sulfur from the fuel to form Sulfuric acid, which has a serious corrosive effect.

Hot air is fed to the combustion chamber to mix with the fuel. Again a critical balance is necessary. If too little air is introduced, the fuel cannot burn completely, with waste and decreased efficiency. If too much air is introduced, work is wasted in pumping excess air, and also the excess air tends to cool the combustion chamber by absorbing heat released in the combustion process. All these processes must be monitored and regulated carefully for maximum efficiency.

We turn now to the steam generation system. Heat from combustion causes steam to form in the primary steam generation coils. The steam bubbles rise into the steam drum, where steam is temporarily stored. From the steam drum the steam goes to a superheater, which uses combustion heat to further heat the steam well above its previous temperature. This feature, with its important increase in steam temperature, is responsible for an important increase in steam engine efficiency. The steam next flows to the turbine. In modern plants the turbine has both high and low pressure stages. The steam flows first into the high pressure turbine, passing from one stage to the next past the turbine blades with a decrease in temperature and pressure at each stage. From the high pressure turbine the steam is fed back to the boiler where it is reheated, nearly to its former temperature, though at a much lower pressure. It then goes to the low pressure turbine and then to the condenser. The feedwater (condensed steam) goes back toward the boiler. But it is now quite cold and it will require a great deal of heat to generate steam from it. Also the boiler would be seriously strained by the addition of this relatively cold water.

Hence, it is desirable to heat the feedwater before returning it to the boiler. To accomplish this, a small amount of steam is bled off from successive turbine stages. Steam is taken from increasingly hotter stages as the feedwater gets hotter. Finally the feedwater is fed into the economizer, which is another set of tubes fairly high in the combustion chamber where the temperature has decreased somewhat. From here the feedwater goes to the boiler water drum and the primary steam generation coils where we started.

Example 5-2

Find the rate at which steam flows in the Centralia coal plant mentioned in Example 5-1.

Each of two units at Centralia operates at 700,000 KW with an efficiency of 40 percent. This means that the equivalent power supplied by coal is:

$$\frac{700,000}{0.4} = 1,750,000 \text{ KW}$$

3,413 BTU is equivalent to 1 KWH. From this we see that 1 KW is equivalent to 56.9 BTU/min. Thus the heat supply rate from the fuel is:

$$1,750,000 \times 56.9 = 10^8 \text{ BTU/min.}$$

Reference to steam tables shows that for the temperature and pressure of the Centralia plant it takes about 1,400 BTU to heat each pound of water enough to cause the required steam to form. With this value and the heat supply rate we can determine the steam circulation rate for each unit.

$$\frac{10^8 \text{ BTU/min.}}{1,400 \text{ BTU/lb.}} = 71,500 \text{ lbs./minute} = 1,190 \frac{\text{lbs.}}{\text{sec.}}$$

Clearly the rate at which steam circulates through a major power plant is very large.

The final topic we shall discuss in this section is the condenser. It is the task of the condenser to cool the steam and condense it to liquid water since, as we saw earlier, the amount of work we can obtain from a given heat source depends on the reservoir, or reference temperature level available. The condenser decreases this reference temperature level. Even with a condenser, the efficiency of a modern fossil-fuel plant is around 40 percent and that of a nuclear plant around 30 percent. Of the 60 percent of the fuel heat which is lost in a fossil-fuel plant the great majority (50 to 55 percent) is carried away by the condenser. The remainder is lost in stack gases, leakage in the system, and mechanical losses. Thus the condenser must be able to carry away about 55 percent of the heat produced in the combustion process.

The condenser gets rid of this 55 percent of the fuel heat in fossil-fuel plants (or about 65 percent in atomic plants) by circulating "cool" water through pipes in the condenser chamber. These pipes with the cool water act as a sort of reverse radiator, cooling the condenser chamber, which leads to condensation of the steam. A condenser is shown in Figure 5-7.

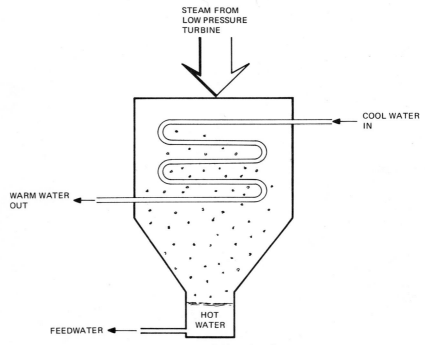

FIGURE 5—7. Steam Condenser.

Cooling water for condensers can come from a number of sources. Some are very cool and some not so cool. Common sources of cooling water include the oceans, rivers, lakes, man-made ponds, spray ponds, and cooling towers. Because the problem of disposing of a huge amount of heat in cooling waters is so critical, we shall devote Chapter 12 to the general problem of waste heat. Before we leave condensers, however, let us take a look at how much cooling water we will need.

Example 5–3

What is the rate at which cooling water must flow in the Centralia plant of Example 5-2? Assume that the cooling water temperature rises 10°.

We need to recall that 1 BTU raises 1 lb. of water by 1° F. The heat which must be dissipated by the condenser is 0.55×10^8 BTU/min. which thus requires a cooling water flow rate of:

$$\frac{0.55 \times 10^8 \text{ BTU/min.}}{10 \text{ BTU/lb.}} = 5.5 \times 10^6 \ \frac{\text{lb.}}{\text{min.}}$$

Since water weighs 62.4 lbs./cu. ft. we can express the water flow rate as:

$$\frac{5.5 \times 10^6 \text{ lb./min.}}{62.4 \text{ lbs./cu. ft.}} = 8.82 \times 10^4 \text{ cu. ft./min.}$$
$$= 1,470 \text{ cu. ft./sec.}$$

We shall leave to Chapter 12 the very important question of how we can dissipate the energy which the condenser's cooling water has picked up going through the condenser.

5.3 THE CONSTRAINTS OF THERMODYNAMICS

The Second Law of Thermodynamics states that it is impossible to construct an engine which, when operating through a complete cycle, will convert all of the heat supplied to it into work.

In this section we discuss this law in a little more detail, particularly as it applies to steam power plants. The most important general implication of this law is that, since the steam power plant is a heat engine, we cannot convert all of the energy released as heat in the combustion process into work. That is, the efficiency of the system must be less than 100 percent. As we saw in the previous section it is typically close to 40 percent for a fossil-fuel plant.

In the previous section we followed the steam through its path in the system, from boiler water into steam, out to the turbine, through the condenser and back into the boiler. This process can be referred to as a steam cycle. In the field of thermodynamics a number of ways have developed for studying this cycle. One is a plot of the pressure of the medium versus its volume, called a steam power pressure-volume (P-V) curve. A typical P-V curve is shown in Figure 5-8. The cycle shown here is a so-called ideal Rankine Cycle, which is commonly used as a standard of reference in analyzing the performance of steam power plants.

To assist in the explanation of the steam cycle, we relate it to a simple piston steam engine, also shown in Figure 5-8. Although this device looks different from a turbine or other steam engines we have discussed, its operation involves all of the same features of heat addition, steam expansion, heat subtraction, and recovery of feedwater.

The line AB in the diagram stands for the heating of the water in the boiler. There is very little volume change here. In this range, heat comes from an external source, the fuel in use. BC represents the process of evaporation or steam formation. At point C steam is released into the cylinder with the piston as far to the left as possible. The line CD represents expansion of the steam, and the corresponding decrease in pressure. As the steam expands, it pushes the piston to the right, doing the work for which the engine is intended. Now we are at a most interesting point (D) in the cycle. Suppose we chose to leave the now-expanded steam in the cylinder and repeat the process. We would have to compress that steam back to its original pressure and volume, that is, retrace the DC path. But to do so would require the same amount of work just

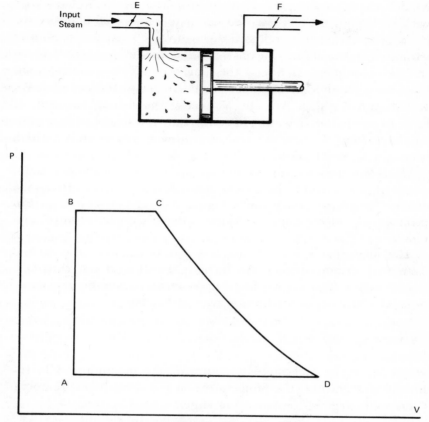

FIGURE 5–8.   Rankine Cycle.

obtained in the expansion stroke,* leaving us with a net zero external work. Clearly this will not do. We have a solution. At point
D we open a valve (F) which releases the steam from the cylinder.
Hence the piston moves back to the left without having to recompress the steam.

Now, what shall we do with the steam released at point D? We
discussed this in the previous section and decided to run it into a

*Actually, a little *more* work would be required to overcome friction.

condenser which will cool it and feed it back to the boiler. Faced with the argument as presented here, students often ask why we cannot pump the steam released through valve F back to the boiler without condensing it. One answer is that any way we chose to do this would require essentially the same amount of work as would have been required to recompress the steam in the cylinder. We would gain nothing. We must condense the steam and pump it back to the boiler as water. In condensing the steam we lose heat, which is released to the external environment in some form. There is no way to avoid this loss.

We accept this loss, undesirable as it is, because of the net useful work obtained from the steam turbine. The turbine drives the electric generator, as shown in Figure 2-3. The generator in turn produces the electric energy which is the end objective of the system.

The unavoidable loss of heat is a manifestation of the Second Law of Thermodynamics. An obvious question arises. How efficient can a heat engine be? For all heat cycles the maximum possible efficiency is the Carnot cycle efficiency:*

$$e = \frac{T_1 - T_2}{T_1}$$

where T_1 is the highest temperature of the medium and T_2 the lowest temperature (the temperature in the condenser). Temperatures are in degrees Rankine (Fahrenheit + 460).

Example 5-4

What is the Carnot efficiency of a modern steam power plant operating at about $1,000°$ F? Assume that the condenser operates at about $100°$ F.

$$= \frac{1,460 - 560}{1,460} = \frac{900}{1,460} = 61.6\%$$

*Sadi Carnot (1796–1832), a French physicist, was one of the first persons to recognize the true nature of heat. He is best known for the ideal efficiency–temperature relationship that bears his name.

This is somewhat higher than the 40 percent suggested earlier. The Carnot efficiency is an ideal limit and many practical problems keep systems from reaching this limit.

5.4 COMBUSTION

Combustion is a rapid chemical reaction in which molecules of one substance combine with molecules of oxygen with an associated release of heat. Fuels used in the combustion process can be solid (such as coal or wood), liquid (oil or derivatives of oil, such as gasoline) or gas (natural gas or manufactured gas).

The principal chemical elements which appear in these fuels, either in elemental form or in compounds, are Carbon, Hydrogen, and Sulfur. The heat obtained from burning the Sulfur is generally not very significant because the Sulfur content is usually low in percentage. However, the Sulfur does play an important role in the production of Sulfur compounds, which are corrosive both in the power plant and outside it and are therefore environmentally undesirable.

Combustion takes place whenever any of these fuels is heated in the presence of Oxygen to a sufficiently high temperature, called the *ignition temperature*. The latter may be as low as 470° F for Sulfur or as high as about 1,200° F for Methane (CH_4), an important constituent of natural gas. The ignition temperature is just high enough to sustain a chemical chain reaction. The heat released by one reaction must be enough to increase the energy of neighboring molecules so that they in turn react chemically with other fuel elements or Oxygen, with the release of more heat. So the chain reaction proceeds. There are a great many chemical reactions which may occur. A few examples are:

Carbon to Carbon Dioxide
and heat $\quad\quad\quad\quad\quad\quad\quad C + O_2 \rightarrow CO_2 + Q$

Hydrogen to water and heat $\quad 2H_2 + O_2 \rightarrow 2H_2O + Q$

Methane to Carbon Dioxide
and heat $\quad\quad\quad\quad\quad\quad\quad CH_4 + 2O_2 \rightarrow CO_2 + 2H_2O + Q$

Sulfur to Sulfur Dioxide
and heat $\quad\quad\quad\quad\quad\quad\quad S + O_2 \rightarrow SO_2 + Q$

In all these reactions, Q stands for a quantity of heat released when the reaction occurs.

The reaction itself is a process in which atoms form new physical alignments and bonds resulting in new compounds, with less net energy than the pre-reaction substances. The net release of heat energy produces the steam required in the steam cycle.

In the chemical reactions above and in all conventional heat-producing reactions, the Carbon, Hydrogen, or Sulfur comes from fuels. It is also necessary, of course, to provide a source of Oxygen to the combustion chamber. Pure Oxygen could be used, but it is expensive to obtain, so we use ordinary air. The primary constituents of air are:

| Constituents of Air | Approximate Concentration | |
	By Volume	By Weight
Nitrogen	78%	76%
Oxygen	21%	23%
Argon	1%	1%
Others	Very small	Very small

Air is clearly an excellent source of the oxygen needed for combustion. It also contains nitrogen, which can combine with oxygen to form undesirable compounds, as we shall see. As we mentioned in Section 5.2, it is important that the amount of air combined with the fuel be just right. Too little air leads to incomplete combustion and waste of fuel. Too much air absorbs heat.

After the combustion has occurred, the resulting waste products are released to the atmosphere through stacks. The products released depend on which fuels are used. In general the important products are various compounds of Nitrogen, Carbon, Hydrogen, Oxygen and Sulfur, including principally Carbon Dioxide, Sulfur Dioxide, Water, and Nitrogen Dioxide.

In addition to the constituents important in chemical reactions, fuels also contain a certain amount of inert matter which appears as a combustion by-product in the form of ash or *particulate matter*. This is particularly true of coal and to a much lesser de-

FIGURE 5—9. Electrostatic precipitators. A bank of eight electrostatic precipitators is shown under construction at the coal–burning plant at Centralia, Washington. These huge devices, which dwarf the workmen around them, remove over 99% of the particulate from the stack gases. Cost of these precipitators was $47 million, or about 18% of the total plant cost at this installation. (*Courtesy Pacific Power and Light Co.*)

gree, oil. As waste, this matter appears either as an "ash heap" at the bottom of a combustion chamber, or as fine particulate matter which is vented to the atmosphere with combustion gases.

5.5 THE ENVIRONMENT

Fossil–fuel power plants have a significant impact on the environment in three ways. First, they release about 60 percent of the heat available from their fuel as waste heat to the environment.

Most of this is released from the condenser to cooling water. Second, they release miscellaneous undesired combustion by-products, in quantity and kind depending on the fuel burned and the heat of the reaction. An environmental advantage of fossil-fuel plants is that they can often be located much nearer to load centers (such as cities) than hydro or nuclear power plants. This decreases the need for unsightly transmission lines which dissipate some of the power transmitted.

We discuss the problem of waste heat, its effect on the environment, and ways of dissipating it, in detail in Chapter 12. Here we shall outline the problem briefly. Cooling water has usually come from rivers, lakes or the oceans. The most common system is "once-through" cooling, in which the cooling water is used once, and then returned to its source. The heated water causes a local increase in water temperature. The heat is then dissipated through mixing and evaporation. The effects of the local heating on the water environment are the subject of much controversy. Some environmentalists believe that this added heat has been the cause of major environmental changes. They often refer to the effect as "thermal pollution." Others claim that there is little or no evidence of damage, and perhaps even some evidence of improved conditions due to heating.(2) Some have suggested that the effect might be called "thermal enhancement" or even "thermal enrichment." One neutral compromise name, which seems reasonable for the time, is "thermal addition." We know that heat is added. We are not certain just what effect this heat has on the biological inhabitants of the cooling water source.

There are two major alternatives to once-through cooling. They are *cooling ponds* and *cooling towers.* Cooling ponds are simply holding ponds in which the heated water is allowed to evaporate. Cooling towers are devices which create a rain of hot condenser water, maximizing the contact of warm or hot water with cool air. In one form, called a wet tower, the cooling air flows past the hot water itself. In a second form, called a dry tower, the hot condenser water remains in radiating pipes, with the cooling air acting on the pipes, but not the water.

The second environmental problem relating to fossil-fuel plants is the problem of the by-products of combustion. Coal-burning plants emit huge amounts of smoke or particulate matter. This material can cover a wide neighborhood of the plant with dust. It is now possible to fit coal-burning plants with electrostatic precipitators which can remove from 99 percent to 99.5 percent of the particulate matter from the stack exhaust when they work properly. Such devices are fairly expensive. They are a good example of how the costs of "cleaning up the environment" must be paid for by the user of the product—electric power in this case. For example, the electrostatic precipitators at the 1,400 MW Centralia coal-fueled plant mentioned earlier cost just over $47,000,000 or about 18 percent of the total plant cost.(3) This cost must be passed on to the consumer. For a number of reasons, however, the increase in the customer's bill will be much less than 18 percent.

The principle gaseous pollutants released from the stack are Carbon Dioxide, Sulfur Dioxide and Nitrogen Dioxide.

The effects of carbon dioxide are not clearly understood. It is required by plant-life for photosynthesis, but an excess may lead to undesirable effects, such as the "greenhouse" effect: a drastic build-up of CO_2 will raise the earth's temperature. To date there is not sufficient evidence to accept or reject the hypothesis of the greenhouse effect.

Sulfur Dioxide is released principally from coal and oil, with amounts highly dependent on the original content of Sulfur; this can vary significantly. Sulfur Dioxide is very heavily studied as a pollutant. It is believed to have important detrimental effects on the lungs of human beings, particularly persons with other respiratory problems. It also has certain detrimental effects on some vegetation. At this time we are not certain just what effects Sulfur Dioxide has on the environment. One author(4) has recently suggested that such emissions may actually be beneficial under certain circumstances, particularly away from urban areas.

The major attack on the Sulfur Dioxide problem at this time is for government agencies to require that plants burn low-sulfur coal and oil, near one percent in content rather than three or four percent. Another partial solution is to use very tall stacks, over

1,000 feet in height. These have the effect of dispersing the undesirable gases. One disadvantage is that they are unsightly. Tall stacks are generally felt to be only a short-term solution. Other approaches are to remove the Sulfur either from the fuel before combustion or from the combustion gases after combustion. Both approaches face technical and economic difficulties at this time. Sulfur removal systems, either before or after combustion, may add as much as $5 to $50 per KW to plant cost.

Nitrogen Dioxide is perhaps the major pollutant released from plants burning natural gas. The combustion process originally forms Nitrous Oxide (NO) as a result of the combination of Nitrogen and Oxygen from the air. The amount of NO formed increases with the flame temperature (at least 2,900° F is required for appreciable NO formation), and the time the combustion gases remain in the high temperature zone. The amount of excess oxygen also affects NO formation, but in conflicting ways. The amount of NO can be reduced significantly by adjusting the three factors listed above.

After the NO is released to the air it combines with Oxygen to form Nitrogen Dioxide (NO_2) which is a yellow-brown gas. It is toxic to animal or plant life if it exists in sufficient concentrations. It also can be highly visible as an important component of "smog."

5.6 THE ECONOMICS OF FOSSIL-FUEL PLANTS

In this section we review the economic considerations peculiar to fossil-fuel plants. The two basic cost factors, fixed costs and operation and maintenance costs, apply here as in hydroelectric plants. Also, the calculation of cost per KWH is as before. The major difference in the economic analysis of fossil-fuel plants, compared to hydro, is that fossil-fuel plants usually have somewhat lower fixed costs, but much greater operation and maintenance costs. The greater operation costs are due largely to the cost of buying and handling the fuel and handling the waste products.

Cost per kilowatt for construction of a fossil-fuel plant varies greatly with such factors as the location, size of the units, type of fuel, operating temperature and pressure, and others. A recent Federal Power Commission document(5) gives estimated 1968 investment costs which range from $95 to $145 per kilowatt, with the average near $125/KW. (These costs are rising because of inflation and the need for new pollution–control devices.) These figures do not include the cost of cooling towers, which are being required in an increasing number of sites.

The estimated service life for modern high–pressure, high–temperature thermal units is about 30 years. This is the time period generally used in amortization calculations.

Other fixed charges include insurance and taxes, and a fund for interim replacements, that is, replacement of components which wear out during the 30–year life of the plant.

Also listed under annual capacity costs are carrying cost of maintaining a fuel inventory and operation and maintenance costs. Operation costs here include the fuel necessary for start–up or to keep the plant on the line, though not producing power. These costs are considered to be a part of fixed annual capacity cost.

The cost of energy used during generation periods is accounted for as a variable operating cost, dependent on the power demand on the plant.

These principles are illustrated by an example adapted from an FPC report(5) shown in Table 5-2. This example is for a 1,600 megawatt, coal-fired plant with a plant factor of 55 percent, fuel cost of $.24 per million BTU, and a heat rate of 8,800 BTU/KWH. The heat rate is the number of BTU which must be supplied to obtain one KWH of electric energy.

The reader will note that the cost of money is accounted here in a way different from that in Chapter 4. Cost of money is indicated as 7 percent and depreciation as 1.06 percent. This assumes that the borrower pays the lender 7 percent of the total capital each year for interest only. In addition the company puts 1.06 percent of the capital into a sinking fund. In 30 years this accumulated investment equals 100 percent of the capital, which is then returned to the lender. This is the same as if the borrower had

TABLE 5-2.

Economic Analysis for a 1,600 MW Coal-Fired Plant

Plant Factor: 55%; Fuel Cost: $.24/million BTU; Heat Rate: 8,800 $\dfrac{BTU}{KWH}$

		Dollars Per Net Kilowatt
A. Plant Investment		$125.00
B. Annual Capacity Cost		
I. Fixed Charges	Percent	
a. Cost of Money	7.00	
b. Depreciation	1.06	
c. Interim Replacements	0.35	
d. Insurance	0.25	
e. Taxes	5.00	
Total Fixed Charges	13.66	*17.08
II. Annual Carrying Cost of Fuel Inventory		0.23
III. Fixed Operating Costs		
a. Fuel		1.01
b. Operation and Maintenance		1.05
c. Administrative and General		0.63
Total Fixed Operating Costs		2.69
Total Annual Capacity Costs		20.00
C. Energy—Variable Operating Costs		Mills/net KWH
a. Energy Fuel		1.90
b. Operation and Maintenance		0.12
		2.02

*$17.08 is obtained by multiplying 0.1366 by $125.

paid 8.06 percent interest to the lender, in which case he would have paid both capital and interest at the same time. Hence, while the bookkeeping is different in this example, the effect is the same as it is in the example in Chapter 4. The reader can easily check this by noting in Table 4-1 that the yearly payment percentage, r, for a seven percent interest rate over 30 years is 8.1 percent. A more accurate table gives this as 8.069 percent.

The total annual capacity cost in Table 5-2 is seen to be $20. To find the capacity cost per KWH we divide by the number of hours of use, which is 8,760 (hours per year) times 0.55 (the assumed plant factor).

$$\text{Fixed Energy Cost} = \frac{\$20}{8,760 \times 0.55} = \$0.00415 = 4.15 \frac{\text{mills}}{\text{KWH}}$$

It is seen that fuel costs and operation and maintenance show up a second time in Table 5-2 under "Energy-Variable Operating Costs." This section (C) represents costs for fuel expended in generating power, whereas section B accounts for fuel used in start-up and stand-by operations. The operation and maintenance accounting under fixed costs is that which is necessary to maintain the plant whether it produces power or not. The same item under section C accounts for special maintenance costs which depend on how much energy is produced by the plant.

Total energy costs are:

Fixed Energy Cost $4.15 \frac{\text{mills}}{\text{KWH}}$

Variable Energy Cost $2.02 \frac{\text{mills}}{\text{KWH}}$

Total Energy Cost $6.17 \frac{\text{mills}}{\text{KWH}}$

REFERENCES FOR CHAPTER 5

1. J.D. Storer, *A Simple History of the Steam Engine*, John Baker Publishers, London, 1969.
2. J.R. Adams, "Thermal Effects of Electric Power Plants," Pacific Gas and Electric Co., 1970.
3. D.G. Van Hersett, "Capital Requirements to Meet Environmental Criterion for Large Thermal Plants," Second Annual Thermal Power Conference, Washington State University, Pullman, Washington, October 8, 1971. (Supplemented by private communication subsequently.)

4. F.F. Ross, "What Sulfur Dioxide Problem?" *Combustion,* September, 1971, pp. 5-11.
5. Federal Power Commission, "Hydroelectric Power Evaluation," U.S. Government Printing Office, 1968.

ADDITIONAL READING FOR CHAPTER 5

1. *Steam—Its Generation and Use,* The Babcock and Wilcox Co., New York, 1972.

This is a very complete and detailed discussion of combustion systems written at a level such that the layman can understand much of the material and get from it a good idea of how such systems work.
2. J.D. Storer, *A Simple History of the Steam Engine,* John Baker Publishers, London, 1969.

This is a very readable basic history of steam engines. The reader can learn much about the Industrial Revolution and about how man has harnessed steam.
3. H.C. Van Ness, *Understanding Thermodynamics,* McGraw-Hill Book Co., New York, 1969.

This is a very good short development of some of the simple thermodynamic concepts introduced in Chapter 5. Much of it can be read by persons lacking a background in physics.
4. H.F. Lund, *Industrial Pollution Control Handbook,* McGraw-Hill Book Co., New York, 1971.

This reference book provides an extensive and up-to-date review of pollution-control techniques and equipment and related control legislation. It is encyclopedic and technical, but many parts of it have value to the non-technical reader.
5. *The Economic Impact of Pollution Control,* Council on Environmental Quality, U.S. Government Printing Office, March, 1972.

This is a summary of recent studies on the cost of providing pollution control in a number of industries. It focuses on the problem of adding control costs to production costs, and it explores the implications of such actions.
6. J.C. Esposito, *Vanishing Air,* Grossman Publishers, New York, 1970.

A Ralph Nader study group produced this review of the causes, the threat, and some of the politics related to the problems of air pollution.

7. J.C. Redmond, J.C. Cook, and A.A.J. Hoffman, *Cleaning the Air: The Impact of the Clean Air Act on Technology*, IEEE Press, New York, 1971.

This is an excellent collection of engineering papers on air-pollution abatement.

PROBLEMS FOR CHAPTER 5

General Problems

5.1. Find the volume of natural gas and the number of barrels of oil consumed each minute if the steam power plant at Centralia, Washington, mentioned in Example 4–1 were run on one of these fuels instead of coal.

5.2. How many acre-feet of coal are used each day by the Centralia coal–fired plant? Assume coal weighs 85 lbs. per cu. ft. (9.5 acre–feet)

5.3. Assume a steam power plant has a condenser temperature of 100° F. Plot the Carnot efficiency as a function of the maximum steam temperature T_1, for $100° F \leqslant T_1 \leqslant 1500° F$.

5.4. Write a short description of the combustion process in which Carbon and Oxygen are combined to produce Carbon Monoxide, Carbon Dioxide, and heat. Discuss the effects of heat, pressure and other important parameters.

5.5. Write a short paper on the effects of Nitrogen Oxides on plant and animal life.

5.6. Repeat the analysis of Table 5–2 for the following new conditions. Cost of money 8 percent. Plant Factor 60 percent. Fuel cost 2.03 mills/net KWH.

5.7. How many cubic feet of natural gas must be burned in a plant with efficiency 40 percent to run a 100 watt light bulb for eight hours?

5.8. Suppose that the addition of various pollution–control devices increases the capital cost of the plant of Table 5–2 by $25 per KW, and the variable operating costs by 0.2 mills/ KWH. What is the new total energy cost?

Advanced Mathematical Problems

5.9. *Heat rate* is the number of BTU which must be supplied to obtain one KWH of electric energy. Obtain a relation between plant efficiency and heat rate. Plot heat rate versus plant efficiency. On the same graph plot heat release to the condenser in BTU/KWH as a function of efficiency, assuming that 5 percent of the input energy is lost in the stacks or other non–condenser loss forms.

5.10. Consider a 700 KW oil–fired plant with a heat rate of 9,000 BTU/KWH. Assume that 1,300 BTU are required to heat each pound of water to the desired steam temperature and pressure. Find the steam circulation rate in pounds per second.

5.11. Consider Example 5–3. Plot the speed of the water flowing into and out of the condenser as a function of the condenser pipe diameter, for a diameter of 0 to 15 feet. What considerations affect the selection of the pipe diameter?

5.12. A recent issue of *Power Engineering* (June, 1972) suggests a fossil–fuel plant capital cost average range of:

Year	1965–'67	1970–'71	1975	1980
$/KW	100	210	260	360

Using these figures and assuming the remaining data are the same as those of Table 5–2, find and plot as a function of time the total energy costs. How realistic is it to ignore rises in other costs?

Advanced Study Problems

5.13. One of the most controversial present fossil–fuel plant sites is at the Four Corners area in the Southwest. Obtain a number of references on this site. Discuss the reason for locating plants at this site, the markets for the power, and the environmental problems related to plants at this location.

5.14. Make a detailed study of methods presently available or proposed for removing Sulfur from fossil fuels before combustion, or removing Sulfur compounds after combustion. Find the fixed costs and the operating costs of these methods.

5.15. Write a paper on the physiological effects of Sulfur and Nitrogen compounds on man.

5.16. Discuss a way of deciding the maximum amount which ought to be spent on air pollution control.

6

ATOMIC FISSION POWER PLANTS

Some recent work by E. Fermi and L. Szilard, which has been communicated to me in manuscript, leads me to expect that the element Uranium may be turned into a new and important source of energy in the immediate future.

> Albert Einstein
> (Letter to President Franklin D. Roosevelt, August 9, 1939)

There is little question that in the field of electric energy generation, atomic fission power plants are the wave of the immediate future. In 1973 about three to four percent of the electric energy consumed comes from atomic power plants. However, about half of the plants proposed or under construction are atomic, and it has been estimated that by the year 2000 about half of our electric energy will come from atomic plants.

Like the fossil-fuel plants discussed in Chapter 5, atomic fission power plants are thermal plants. They heat water to steam which passes through a steam turbine which in turn drives an electric generator. The steam cycle is essentially the same as for fossil-fuel plants except for the source of heat, which is a controlled nuclear reaction. In this chapter we shall emphasize that distinguishing feature, with little or no discussion of the basic steam cycle (Chapter 5) or the condenser cooling system (Chapters 5 and 12).

We start with an introductory look at nuclear physics, showing how energy can be obtained in a controlled way from certain nuclear, as opposed to chemical (fossil-fuel), chain reactions. We consider next how the atomic fuel is assembled in the "reactor

FIGURE 6–1. Diablo Canyon nuclear power plant. When completed in 1975 or 1976, each of the two Diablo Canyon nuclear units will produce 1,060 MW of power. The original cost estimate for the plant was $400 million, or about $200 per kilowatt. But escalating costs will raise that figure quite significantly by the time construction is complete. In late 1973 a geological fault was discovered nearby; this fault may jeopardize the future of the plant. See page 148. (*Courtesy Pacific Gas and Electric Co.*)

core." We then look at a number of ways of building atomic reactors at this time. This leads us to what many people hope and
expect will be the reactor of the future, the breeder reactor. Next
we discuss the fabrication of the atomic fuel. We review here
the major environmental problems which must be faced as atomic
power develops. Few environmental questions are so difficult to
evaluate, and few are discussed as intently as are those relating to
atomic power. At the end of the chapter, then, we take up certain
economic problems peculiar to the atomic power industry.

6.1 A SUMMARY OF NUCLEAR PHYSICS

Atoms are the basic building blocks of the things of nature.
Atoms consist of a central nucleus containing protons and neutrons, surrounded by rapidly moving light electrons. This structure
is shown in Figure 6-2. The protons in the nucleus have a positive

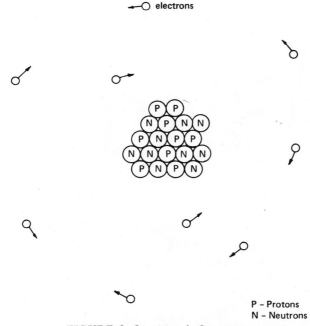

FIGURE 6—2. Atomic Structure.

electric charge. The neutrons have no charge. The number of negatively–charged electrons surrounding the nucleus of a neutral atom (that is, one that has not been *ionized*), just equals the number of protons. Since both protons and electrons have the same unit charge, though it is of opposite sign in the two, the net charge of the atom is zero. At this point the atom is said to be electrically neutral—not ionized. The atom pictured in Figure 6-2 is enormously distorted in scale; the nucleus of a real atom is extremely small, with dimensions on the order of one trillionth (10^{-12}) of a centimeter; the centimeter is about 0.4 inches. The electrons surround the nucleus at distances about 10,000 times the dimension of the nucleus. Therefore, for the nucleus shown in Figure 6-2, the electrons should be drawn far beyond the edges of the paper.

Elements, of which there are over 100 in nature, are characterized by the number of protons in their nucleus. This number is called the *atomic number* of the atom. For example, the element Hydrogen has one proton; Carbon has six; Silicon has 14, etc. The element Uranium has 92. Their atomic numbers are 1, 6, 14, and 92 respectively.

It is possible for some atoms of any one element to have different numbers of neutrons in their nucleus, though of course the number of protons must be the same or they would constitute a different element. Atoms having the same atomic number (protons) but different numbers of neutrons are called *isotopes*. Isotopes are differentiated by their *mass number*, which is the sum of protons and neutrons in the nucleus. For example, there are 14 isotopes of Uranium, three of which occur naturally. These three have mass numbers 234, 235, and 238. Their symbols are:

$$_{92}U^{234}$$

$$_{92}U^{235}$$

$$_{92}U^{238}$$

The lower left hand number is the atomic number, the upper right number is the mass number. The U is the letter symbol for Uranium. Sometimes the lower left number is not used, since it repeats the same information as the letter symbol.

Uranium is of particular interest to us because it is the most important element used in nuclear fission energy generation today. The three isotopes mentioned above occur naturally in the following percentages:

Isotope	Percent (by Weight)
U^{234}	0.006
U^{235}	0.711
U^{238}	99.283

Atoms can release energy in a process called *fissioning*, which can occur under certain limited circumstances. If the right isotope of the right kind of element is struck by a free neutron of the right speed (energy), the isotope can break up into two new elements, with the release of energy, and with the release of one or more additional particles or forms of radiation. It is the additional energy obtained in the fission process which can be harnessed in nuclear reactors. In order to obtain a significant amount of energy it is necessary to have the right materials under the proper conditions to set up a nuclear chain reaction, analogous to the chemical chain reaction required for combustion. In an atomic device, a chain reaction exists when sufficient neutrons are released and allowed to interact with other atomic nuclei to continue the fission—neutron release—new fission process or "chain."

There are hundreds of isotopes in nature, but only a few of them can fission. Some of the most important are

$$_{90}Th^{232} \quad \underline{_{92}U^{233}} \quad \underline{_{92}U^{235}} \quad _{92}U^{238} \quad \underline{_{92}Pu^{239}} \quad _{94}Pu^{240} \quad \underline{_{94}Pu^{241}}$$

The four underlined isotopes have a special property. They can be fissioned by neutrons of any energy, down to zero. These isotopes are said to be *fissile*.

The three isotopes which are not underlined are said to be *fertile*. They can be fissioned only by neutrons with energies above about one million electron volts, written 1 MeV. (An electron volt is a measure of energy which an electron acquires as it passes from one point to another, with a positive electric potential of one volt between the points.) The fertile isotopes are called fertile because, when they capture a high energy neutron, they produce a fissile isotope after a short radioactive decay process.

Of all the hundreds of possible isotopes, *only* Uranium–235 ($_{92}U^{235}$) *is fissile and occurs naturally*. This makes it a most important constituent of most atomic energy systems.

The process of fissioning U^{235} is illustrated in Figure 6–3. A neutron is captured by $_{92}U^{235}$ transforming it to $_{92}U^{236}$, which is highly unstable. In about a millionth of a second or less the $_{92}U^{236}$ fissions (or breaks up) into two large "fission fragments," nearly equal in size, plus one or more neutrons, and gamma and beta radiation (which we will discuss shortly). The fission fragments are isotopes of new elements. There are at least 40 ways in which the original atom can fission, yielding over 80 different fission fragments. There is no way to predict which pair will be produced in a given fission. Furthermore, the resulting fragments are radioactive. That is, they emit radiation of some form, changing to a new isotope in the process. These changes continue in a chain through an average of about five states until they finally reach a stable non–radioactive state. The result of this process is that the fairly simple mix of U^{235} and U^{238} and possibly a very few other constituents originally put into a reactor becomes in time a com-

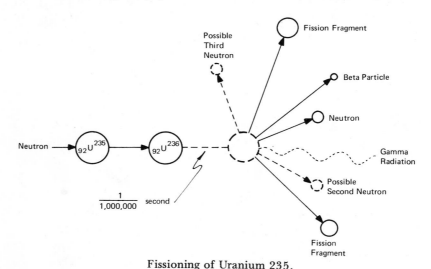

Fissioning of Uranium 235.

plex of about 200 different isotopes. Some are radioactive; some not. Some are highly dangerous, biologically, for a long time; some are not.

Besides the fission fragments, indicated in Figure 6–3, there are also an average of about 2.4 neutrons released per fission for U^{235}. Some of these neutrons are lost to the reaction process, while others interact with other fissile atoms, continuing the chain reaction.

Also, there are beta and gamma radiation releases. We shall return to a more detailed description of radiation shortly, but first we have to talk about the reason for building atomic reactors: energy.

When we add up the masses of all of the fragments and particles resulting from the fission process, we find that it is *less* than the mass of the original $_{92}U^{236}$. The mass which is lost has been changed into energy in accord with the famous relation postulated by Albert Einstein near the turn of this century:

$$E = mc^2 \qquad\qquad\qquad (6\text{-}1)$$

where E is energy (in ergs), m is the mass (in grams) and c is the velocity of light (about 3×10^8 meters/sec.). The energy increase resulting from the fissioning of a $_{92}U^{236}$ atom is about 200 MeV. Most of this energy (about 85 percent) goes into kinetic energy of the fission fragments. The rest is distributed among the other fission products, as shown below:

	Energy (MeV)
Kinetic energy of fission fragments	168
Kinetic energy of fission neutrons	5
Energy of gamma rays	11
Energy of beta particles	7
Total heat energy	191
Neutrino energy (not available as heat)	11
Total	202

The total heat energy is available to produce steam to drive the turbine of the electric power plant.

To see the significance of the release of 191 MeV of energy from an atom of fissile Uranium, let us return briefly to the chemical reaction which takes place in combustion. When an atom of Carbon combines with two atoms of Oxygen to form CO_2, about four eV are released. This is smaller than 191 MeV by a factor of about 50,000,000.

Accounting for the difference in weights, we find that one pound of fissile material can release about as much energy as 1,400 *tons* of coal. When breeder reactors are developed, the ratio of required coal to fissile material will be far greater.

Example 6-1

Assume that the percentage of fissile material (usually U^{235}) in reactor fuel is 2%. About how many tons of reactor fuel must be used each year in a plant equal to the size of the Centralia coal plant discussed in Chapter 5? How many cubic feet does this weight of reactor fuel equal? Assume a plant factor of 80 percent.

In Example 5-1 we saw that the Centralia plant requires 12.25 tons of coal per minute. Thus the coal required per year is:

$$12.25 \times 60 \times 8{,}760 \times 0.8 = 5{,}140{,}000 \text{ tons}$$

Since about 2 percent of the fuel is fissile, we require 1 lb. of Uranium fuel for each $0.02 \times 1{,}400 = 28$ tons of coal. Thus the required Uranium fuel is

$$\frac{5{,}140{,}000}{28} = 184{,}000 \text{ lbs.} = 92 \text{ tons}$$

Since Uranium weighs 1,205 lbs./ft.^{3}, the volume of 184,000 lbs. is:

$$\frac{184{,}000}{1{,}205} = 153 \text{ ft.}^3$$

These numbers assume complete burn-up of U^{235} which does not occur in practice. Typically, a reactor of this size would have fuel replacement weights greater than the above but far less than for coal.

While it is certainly necessary that energy be released in the fission process, the neutrons released are of equal importance, because some of these neutrons will collide with other fissile atoms causing a new fission, with release of more energy and more neutrons. The object is to create a sustained chain reaction analogous to the chemical chain reaction required for combustion. There are a number of factors to consider in establishing and sustaining a safe chain reaction. We discuss some of these in later sections, and leave others to more detailed studies of nuclear reactor theory and practice.

6.2 RADIOACTIVITY

Radioactive materials spontaneously emit one or more of three types of radiation, called *alpha radiation, beta radiation,* and *gamma radiation.* This radiation is, in general, harmful to human beings, animals and plants, at least if the intensity is great enough. Such radiation occurs naturally in a number of forms, coming from the sun, from outer space, and from materials all around us. The world has always known such radiation, and it has always been at least potentially harmful.

Another source of radiation is in the decay of the 200 or so possible byproducts of a nuclear chain reaction. The difference between this radiation and that which the world has always known is that nuclear reactor radiation is far more concentrated, and thus potentially harmful or fatal if sufficient precautions are not taken.

We turn now to the three forms of radiation, considering their natures and their effects on man. An alpha particle consists of two protons and two neutrons. It is the nucleus of a helium atom. It is a heavy particle, capable of penetrating only a short distance into materials, depending on its speed or energy. It usually will not pass through a thin sheet of paper. It is not considered an external threat to the body. It can be damaging if it gets inside the body. The effect of the alpha particle is to cause molecules to ionize. If this happens in a human body, unwanted chemical reactions follow and severe damage may take place.

Beta particles are electrons. They are much lighter than alpha particles. They also have a much greater ability to penetrate matter. A beta particle might be stopped by about one-half inch of aluminum. Again, this depends on the speed or energy of the particle. Beta particles also cause ionization, but, because they are lighter, they usually cause much less ionization than alpha particles. If particles have the same energy they cause the same ionization. We might ask just how much ionization a particle can cause as it moves through a material. Suppose the particle has an energy of five MeV. Since it takes an average of about 34 eV to form an ion pair, about 50,000 pairs can be formed by such a particle.

Gamma rays are essentially x-rays. They are electromagnetic waves, as are radio, television, and light waves, but x-rays are of much higher frequency. They have a very great ability to penetrate matter, requiring several inches of lead for adequate shielding. They cause the production of ion-producing particles, and they therefore can cause serious damage to the body.

It is important that we have a measure of the amount of radiation which the body receives, since this radiation can affect the health of the body. It is not a simple task to develop a universally useful unit because different forms of radiation have different effects.

The first generally accepted unit was the *roentgen*, which uses ionization produced in air as its basis. This unit relates directly only to x-rays or gamma radiation.

A more general measure of radiation is the *rad*. It is the amount of radiation which causes the absorption of 100 ergs of energy per gram of matter. This unit, however, does not take into account the relative intensity of biological damage effectiveness of different forms of radiation.

A generally accepted unit which does consider relative damage effects is the *rem* (roentgen equivalent man).

The next obvious question is how many rems can a man be safely exposed to? This question is complex and controversial, particularly for low-level doses, as we shall see in Section 6.7 on environmental problems. For higher level doses, such as might be experienced from an atomic bomb, or from a very serious radiation accident, immediate effects are easier to predict. Doses from 0 to 25 rems would probably show no observable effect. Doses up to 100 rems would probably cause slight blood damage but little or no externally-observable effects. In the range 100 to 200 rems, "radiation sickness" may occur in some individuals, with vomiting in a few hours, fatigue, loss of appetite, and recovery in a few weeks. From 200 to 600 rems, radiation sickness will be more severe and may also be accompanied by loss of hair. Loss of life in a few weeks or months will occur for many, depending on the exposure and their state of health. More than 600 rems of exposure leads to massive radiation sickness, vomiting within an hour, severe blood change, hemorrhage, infection and loss of hair. Most persons exposed to this level die within two months.

The length of time that a material remains radioactive is usually expressed in terms of its *half-life*. Assume that a material has a certain number of radioactive atoms. These will decay spontaneously and randomly, the atom changing to some new isotope, with the release of some form or forms of radiation. The half-life of the material is defined as the time required for half of the atoms to decay. In one half-life, half of the atoms decay. In the next half-life, half of the remaining atoms decay. This process continues indefinitely. Half-lives of some radioactive isotopes are given below.

Isotope	Half-Life
Carbon-14	5,580 years
Plutonium-239	24,300 years
Plutonium-241	13 years
Uranium-239	23.5 minutes

Example 6-2

If you start with one pound of Carbon-14, how much is left in 27,900 years?

27,900 is 5 half-lives for Carbon. Hence the weight is reduced by $1/2 \cdot 1/2 \cdot 1/2 \cdot 1/2 \cdot 1/2 = (1/2)^5 = 1/32$. Thus the weight is 1/32 lb.

6.3 THE REACTOR CORE

In this section we describe the basic structural form and components of a nuclear reactor. The essential parts of a reactor are the radioactive fuel, a moderating material, control rods, and a means of removing the heat from the reactor and getting it,

FIGURE 6–4. Reactor vessel delivery day. This huge 345–ton reactor vessel is the heart of one of the two 1,060 MW units at Diablo Canyon (Figure 6–1). The pipes on the near end (the bottom of the vessel) are for instrumentation. The large openings near the far end will carry pressurized water to steam generators. The vessel will rest in a concrete vault with walls six feet thick. This is the end of an 8,000 mile journey for the vessel, which was built in Chattanooga. It was barged down the Mississippi and shipped via the Panama Canal to the California site. (*Courtesy Pacific Gas and Electric Co.*)

eventually, to the steam turbine. (A simple core is sketched in Figure 6-5 below.)

The nuclear fuel, which is fabricated as described in Section 6.6, has the form of small cylindrical pellets about an inch in diameter and an inch and a half long. These pellets are placed in long thin cylinders and inserted into the reactor as a unit.

Fast neutrons are not easily captured by $_{92}U^{235}$, so a moderating material is required in the core of the reactor to slow down the neutrons produced by fission and needed for succeeding fissions. Neutrons are produced in fission processes with an energy close to two MeV. This must be reduced to a value near 0.1 eV. Moderators should contain light atoms which do not easily absorb neutrons. Water, hydrocarbons, Beryllium, Carbon, and some other materials make good moderators. Water is the least expensive and often the most convenient moderator. It is common practice to use the same water to carry heat from the atomic core to the steam generator.

Control rods are long cylindrical rods made of some material which absorbs neutrons, such as Boron, or Boron Carbide and

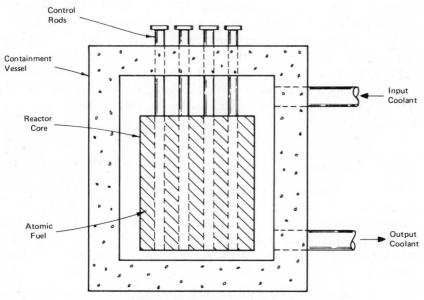

FIGURE 6—5. Nuclear Reactor.

Aluminum. When the control rods are inserted into the core, somewhat like pins in a pincushion, they capture neutrons with the effect of slowing or stopping the chain reaction. Since they can be set in any desired position, they have the effect of controlling the reaction at the desired rate. They are also used as one of the safety mechanisms, since in case of an accident, they will stop the chain reaction when fully inserted into the core.

Heat is removed from the core by a circulating *coolant*. The coolant may be a gas, a liquid, or a molten solid. It is pumped through the core making contact with the hot fuel elements. It then carries the heat out of the core to be used eventually to drive the turbine. There are a number of possible forms for reactors; these are discussed in the following section.

6.4 FORMS OF COMMERCIAL NUCLEAR REACTORS

There are many forms in which a nuclear reactor may be built. We concentrate our attention here on the three basic forms which have been developed for commercial use in the United States. These are the *boiling water reactor* (BWR), the *pressurized water reactor* (PWR), and the *high temperature gas reactor* (HTGR).

In the United States there are five major reactor–building companies and a number of other companies beginning to enter the field. Each company produces a particular form of reactor. The companies are listed below with the type of reactor they build, the number sold, and the total power of the number sold as of September 30, 1970.(1)

Company	Reactor Type	Number	MWe*	Percent
Babcock & Wilcox	PWR	14	11,514	13
Combustion Engineering	PWR	12	10,698	12
General Electric	BWR	41	33,212	38
Gulf General Atomic	HTGR	2	370	1
Westinghouse	PWR	37	29,994	35
Other		3	901	1

*The expression MWe stands for megawatts-electric. It is sometimes used to distinguish electric power out of a plant from thermal power into a plant.

The basic form of the BWR is shown in Figure 6–6a. In this type of reactor steam is produced directly in the reactor and then fed to the turbine. This reactor operates at a pressure of about 1,000 pounds per square inch (psi). The PWR, shown in Figure 6–6b, has two loops, one in which the coolant remains inside the containment vessel. This loop uses a heat exchanger to transfer heat to a second loop which then produces steam for the turbine. Pressures in the first loop are nearly 2,000 psi.

The BWR tends to have the higher radiation releases of the two because the coolant, which flows directly past the fuel elements,

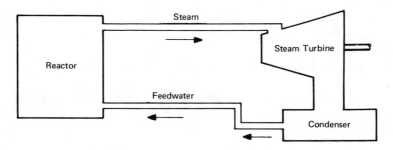

(a) Boiling Water Reactor (BWR)

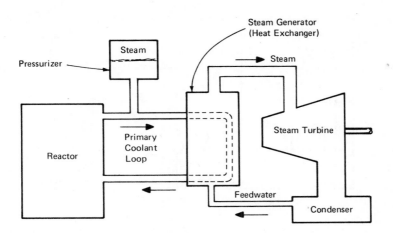

(b) Pressurized Water Reactor (PWR)

Figure 6–6. Forms of Water–Coolant Reactors.

leaves the containment vessel to pass through the turbine. There is greater opportunity for release of small amounts of radioactive contamination from the BWR. (In the case of the PWR, the primary coolant loop is entirely within the containment vessel and the opportunity for release of radiation is reduced.)

On the other hand, the BWR is inherently safer because of the lower pressures encountered in the coolant loop. The BWR has also a safety edge because the boiling water, which itself moderates the chain reaction, becomes a poorer moderator as it turns to steam in case of an accident. This tends to slow down the reaction. Both forms have an efficiency of nearly 32 percent.

In addition to arguments suggested above, there are many other factors tending to favor one or the other of the two systems. Which, then, is the better of the two? There is no clear answer to this. It is not surprising that they share the commercial market almost equally.

The Atomic Energy Commission (AEC) has encouraged the development of other reactor forms. One result has been the High Temperature Gas Reactor (HTGR). The first commercial American HTGR (40 MWe) was completed in 1967 in Peach Bottom, Pennsylvania. The only other HTGR (330 MWe) should be in oper-

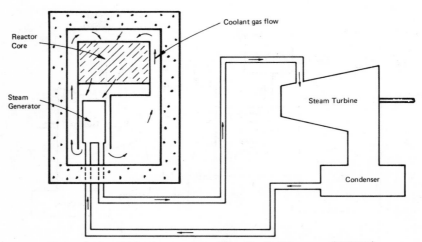

Figure 6—7. High Temperature Gas–Cooled Reactor (HTGR).

ation in 1973 at Fort St. Vrain, Colorado. Recent announcements of building plans for HTGR's add up to over 4,500 MWe. A block diagram of the Fort St. Vrain facility is shown in Figure 6-7. The gas coolant is Helium, which is heated to 1,300–1,500° F. Steam is produced at 1,000° F and 1,450 psi. The high steam temperature leads to an efficiency of about 40 percent, somewhat better than that of the PWR or BWR. Other advantages arise from better fuel utilization and longer periods between re-fueling. A disadvantage is a tendency for the HTGR to be more expensive, particularly for small units, than the PWR or BWR. Larger units are now available and increased interest in the HTGR is evident.

6.5   THE BREEDER REACTOR

The reactors discussed so far have been so-called "burners," in that they burn up or exhaust their supply of U^{235}. They require a moderator to slow down the neutrons so that they can be captured by another U^{235} nucleus for another fission reaction. There are, however, other reactions which can take place. One reaction makes use of U^{238}, which is 140 times as abundant as U^{235}. It is fertile, however, and thus requires high-energy or "fast" neutrons. Such neutrons are present if we remove the moderator used to slow them down in burner reactors. One reaction with fast neutrons is:

$$n + U^{238} \rightarrow \quad U^{239} \quad \rightarrow \quad Np^{239} \quad \rightarrow \quad Pu^{239}$$

fast	23.4 min. half-life	2.3 day half-life	24,000 year half-life

Plutonium-239 ($_{94}Pu^{239}$), as we saw in Section 6.1, is fissile. This reaction, then, breeds an element which can become fissile fuel. If the reactor is to work satisfactorily as a breeder of fuel, it is necessary that more than two neutrons be emitted by the Plutonium as it fissions. If, for example, an average of 2.5 neutrons are produced, one will fission an atom of Pu^{239}, continuing the chain reaction, and 1.5 will convert more fertile U^{238} to fissile Pu^{239}. The result is that more Plutonium is produced than is consumed. The extra Plutonium can eventually be used to fuel other reactors.

It must be emphasized that we are not breeding new fuel from nothing but rather breeding fissile Pu^{239} from fertile U^{238}. Eventually we would use up all of our U^{238}, and also the fuel it has bred. The advantage of the breeder reactor is that it uses U^{238} for its basic fuel material which is 140 times more abundant than U^{235}. This increases the potential life of nuclear reactors as a power source from 30 to 100 years, to perhaps thousands of years.

We cannot use water as a coolant in the breeder because it acts as a neutron moderator. One of the most popular designs presently under study uses liquid Sodium. Other proposed breeders use other coolants, including gas. There are also other possible atomic reactions including one using Thorium-232 to produce Uranium-233.

Much effort is presently underway to produce effective commercial breeders by the 1980's. Small breeders have been built, and some larger models have been built or are under study, particularly in the USSR and in Great Britain. The United States effort centers on research and initial development in the 1970's. Commercial utilization will develop in the 1980's. The form of greatest interest at this time is the Liquid Metal Fast Breeder Reactor (LMFBR).

The major advantage of the breeder is in fuel utilization. It also will be more efficient, with anticipated efficiencies near the 42 percent of modern fossil-fuel plants. One disadvantage is that it will require more core fuel than other reactors; this fuel must be more densely-packed. Present research centers around development of a breeder with sufficient high fast-neutron production and acceptable safety.

6.6 FUEL FABRICATION, REPROCESSING, AND STORAGE

In this section we discuss how fuel is obtained from mines, processed for use in power plants, reprocessed after it is removed from the plant, and sent either to the original processing plants or to waste storage sites. This entire process is summarized in a block diagram in Figure 6-8.

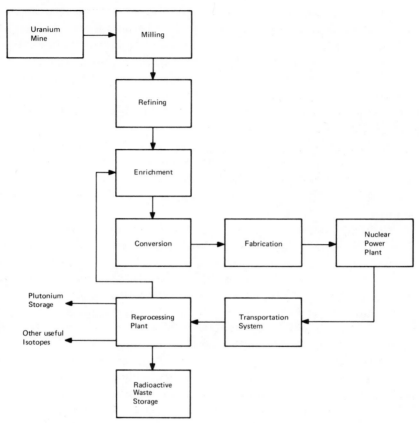

Figure 6—8. Nuclear Fuel Cycle.

The first step is to mine the Uranium from the earth. Some of this is open pit mining and some underground mining. The earth's crust has a large amount of Uranium but most of it is widely dispersed. Principal sources of Uranium are found in the United States, Canada, South Africa, France, Czechoslovakia and Russia. It appears in convenient concentrations in Pitchblende or in Uraninite, an oxide of the metal. Pitchblende has a greater concentration of Uranium but Uraninite is more common. The latter provides about 95 percent of the Uranium in the United States.

Uranium appears in Uraninite in the form of an Oxide of Uranium, U_3O_8, which represents around 0.2 percent of the basic ore. The cost of producing one pound of U_3O_8 is about six to eight dollars at this writing. As U_3O_8 becomes less–easily mined and processed, the cost will go up. An important question in the nuclear power industry revolves about the amount of "eight dollar" Uranium available. As the cost of Uranium goes up, the cost of electric power generated from it will go up, but certainly not proportionally. The cost of Uranium fuel is still a rather small fraction of the cost of delivered electric power. Another factor which complicates the cost analysis is that U_3O_8 will be far more valuable as a fuel in a breeder because utilization will be so much greater. We can afford to spend more to process fuel for breeders than for burners.

We turn now to a more detailed view of the steps involved in the fuel cycle shown in Figure 6–8. Uranium ore is slightly radioactive, and it requires some precautions in its handling. Prolonged breathing of the ore dust can be harmful. It is also necessary to handle the mine "tailings" or refuse with caution. In a recent incident a large number of homes were built in Colorado on ground using mine tailings for fill. The result is a slow release of low–level radiation in the area. (See Section 6.7.)

The U_3O_8 in the ore is concentrated in the milling step by a process which includes crushing, screening, washing and gravitational separation. This concentration step usually takes place near the source of the ore, so that the amount of valuable U_3O_8 which must be transported to other processing stages is small compared with that of the original ore.

The next step in the process is the refining step in which the U_3O_8 is chemically changed into either Uranylnitrate Hexahydrate (called "yellow cake" in the industry) or into Uranium Hexafluoride.

The enrichment step which follow is most important, since it is here that concentration of U-235 (normally 0.7 percent) is increased. This step is also quite difficult, because it involves separating isotopes which are chemically identical. Separation is accomplished by a complex process which takes advantage of the extremely small difference in mass between the isotopes U–235 and U–238. In the United States enrichment is accomplished in a gaseous diffusion process performed at plants at Oak Ridge, Tennessee; Paducah, Kentucky; and Portsmouth, Ohio. These plants are owned by the government but operated by private contractors. (In 1964 they consumed about *45 billion* KWH of electric energy to drive gas–circulating pumps, etc.)

The next step is to convert the enriched Uranium to Uranium Dioxide powder (UO_2). This powder is pulverized and formed into small cylindrical pellets. These are the basic fuel pellets which are to be placed in the core of the reactor. At this point they are only very slightly radioactive and can be held in the hand without serious harm.

In the fabrication process the pellets are placed in long narrow cylindrical rods called *cladding*. The cladding must be strong enough to withstand many years of hot, high–pressure coolant flow. It also must allow neutrons to pass through it fairly easily without serious deterioration. Zircaloy, an alloy of Zirconium, is widely used in water–cooled reactors. Fuel rods are collected in assemblies of perhaps a few hundred, with some rods left empty for control rod insertion. Fuel assembly fabrication is the most expensive processing step, and therefore it is the step in which major cost reduction is being sought.

Next the fuel assemblies go into the nuclear reactor, where they may remain for up to four years. About one fourth of the fuel is replaced each year or so. This process is necessary for a number of reasons. First, a small amount of fuel is burned up, possibly enough to slow or stop the reaction. Second, the fission byproducts tend to capture neutrons. Third, fuel elements and cladding suffer damage and distortion if left in place long enough. There are at least two reasons for sending the *spent fuel* back to a processing plant rather than to *waste* storage, as we shall see shortly.

First let us consider the problem of transportation of spent fuel from the reactor to the processing plant. This is a critical step because we are now moving the highly radioactive fission products produced in the reactor core. Shipping of fuel assemblies is closely controlled by the AEC. Shipping casks must satisfactorily survive the following conditions:

1) Dropping 30 feet onto a hard surface
2) Dropping three feet onto a steel projection six inches in diameter
3) Exposure to fire
4) Submersion in water

The reprocessing plant is the next step, first because unburned fuel can be salvaged and sent back to the processing plant, and second because Plutonium, and other valuable isotopes, can be recovered. Radioactive isotopes which are of no value, and which cannot be released to the environment, are sent to waste disposal sites where they will gradually decay at rates depending on their half-lives.*

The nature of the waste storage system depends on the degree of radioactivity of the waste. Highly radioactive material is stored in welded steel tanks with a capacity of 600,000 gallons. They are housed in concrete vaults, and covered with many feet of earth. Because the continued radioactive decay gives off heat, the liquid gives off heat, and it must be cooled continually. Present AEC plans call for converting wastes to a solid form for more convenient and safer storage.

6.7   ATOMIC POWER AND THE ENVIRONMENT

The major environmental advantage of atomic power plants is that they do not release large amounts of air pollutants (particulates and Oxides of Sulfur, Carbon, and Nitrogen). Fossil-fuel plants also require large quantities of fuel which require many trains, pipelines, or oil tankers.

The problem of condenser-water heat addition (or thermal pollution) is the same in principle for atomic plants as for fossil-fuel plants, because both are thermal plants. However, as we shall see in Chapter 12, atomic power plants release to cooling water about 50 percent more heat than do fossil-fuel plants.

Four environmental problems are related to the use of nuclear power plants. These problems are:

*The curious reader may wish to ponder the implications of this statement. Refer to the list of half-lives in Section 6.2

1) Low-level release of radiation
2) High-level release of radiation
3) Diversion of atomic materials
4) Storage of waste materials

Low-level release of radiation may be expected and permitted by existing regulations, or it may be accidental. Such levels of radiation may be nearly harmless, although there is debate on this question. We refer to levels in the one to 500 millirem range. (A millirem is 1/1000 of a rem.)

Radiation losses occur regularly in nuclear power plants because of leakages from the core or as a result of fuel or waste handling. They are relatively difficult to avoid. Typically, nuclear plants vent to the air or release into waterways radiation which would expose an individual near the plant to about one to 10 mrems per year.

Until about two years ago, or 1971, AEC regulations permitted an exposure of 170 mrems per year to persons in the general population. The issue of the possible danger of such radiation levels was dramatically made public by two Lawrence Radiation Laboratory physicists, John Gofman and Arthur Tamplin.(2) They argued that exposing every person in the United States to these levels would result in an added 32,000 deaths per year from cancer and leukemia. Although the argument was not without flaws, it was no doubt very significant in influencing the AEC decision to reduce permitted exposure levels drastically. Present guidelines call for an exposure of five mrems per year for a person at the boundary of a plant and one mrem for large populations.

Normal background radiation and radiation from medical x-rays contributes about 100 to 200 mrems per year for each person. Hence the present guidelines would seem reasonable at this time. However, these guidelines apply only to operating power plants, and not to processing plants. The latter could probably be controlled without too much added expense, and serious consideration should be given to bringing them under the above guidelines.

The second problem is that of high level release of radiation in the event of a serious accident in the plant or in various stages of processing or transportation. We do not refer here to atomic-bomb-like explosions, which are not possible with the very low concentration of fissile fuel in reactor cores. It *is* physically possible, however, to have serious or even catastrophic releases of radiation as the result of certain types of accidents. One serious and potentially dangerous accident is the "loss of coolant" accident, in which the lines carrying coolant to the reactor burst or coolant pumps fail. This keeps the coolant from carrying heat away from the core. Control rods are immediately inserted into the core to stop the chain reaction. However, if the fuel has been in the core long enough to build up a sizeable inventory of radioactive fission fragments or byproducts, this decaying radiation will contribute significant heat. The result may be a melt-down of the core. There is no material which can really contain this very hot core, and it will melt through containment structures into the ground. Where it will stop or spread to is not known. This possibility is sometimes referred to as the "China syndrome."

Another possible accident involves the rupture of the outer containment vessel when the coolant line breaks. This could permit the release of massive amounts of radiation into the air.

These and other accidents *can* happen. But how probable are they, and how serious would they be? Both questions are essentially impossible to answer. We have had too little experience; the nuclear industry is changing far too rapidly, and the regulating agency, the Atomic Energy Commission, has not published estimates of the probability of a major accident. The result is that we are left without any really good estimate of the safety of nuclear power plants. Every effort is made to make plants "fail-safe"— provide back-up, or emergency, measures to avoid accidents. But the true reliability of these systems and the plant as a whole is not known.

One of the major back-up safety systems in plants is an Emergency Core Cooling System (ECCS). The ECCS would provide an emergency supply of cooling water to the core in the event of an accident. The system has never been tested at full scale. A small scale test in 1970 showed that the water which was to be sprayed into the core was vaporized and blown out through the ruptured containment vessel.(3) Defenders of the system maintain that a computer analysis shows that a full scale test will be successful.

Such a test is planned for 1974 or 1975. The ECCS is not, of course, a single link on which all safety depends. Its failure under the small scale test was nonetheless very disappointing. It tends to support the idea that we are working in an area in which there are many uncertainties. Again, we really don't know how safe nuclear power is.(3)

FIGURE 6–9. Floating nuclear power plants. One of the more imaginative solutions to the problem of siting nuclear power plants is shown here. The Public Service Electric and Gas Company of New Jersey has contracted with Offshore Power Systems to develop a pair of floating nuclear plants in the Atlantic Ocean, three miles off the New Jersey coast. The ocean here is about 40 feet deep. The plants are protected by an enormous breakwater 300 feet wide at the base and 30 feet wide at the top. The breakwater is designed to withstand winds, waves, and storms of much greater magnitude than have been observed in this location. Together, the plants will generate 2,300 MW. Capital cost is estimated at about $750 million, or about $300 per KW. Condenser cooling will raise the temperature of a few acres of the surrounding ocean by about five degrees Farenheit. Power will be transmitted to the land by 345,000 volt cables buried beneath the sea. The cutaway drawing shows the reactor vessel and steam generators on the left, and the steam turbines and electric generators on the right. (*Courtesy Public Service Electric & Gas Co.*)

The threat of earthquakes which might destroy a number of safety systems at once and lead to a serious accident has limited the development of atomic plants in parts of the United States, particularly along the California coast.

The third problem we consider is the danger of diversion or stealing of nuclear fuel, particularly when it is being transported. This problem will become more serious with the introduction of breeder reactors, which will have significant supplies of Plutonium–239. It would be relatively easy to use this material in the construction of a small atomic bomb. Such a device could be used by terrorist groups, small countries, blackmailers, etc. In 10 to 20 years huge amounts of spent fuels will be transported with increasing opportunity for diversion.*

Finally, we consider the problem of the storage of undesired nuclear wastes. Such wastes are radioactive and hot, and some of them must be stored for tens of thousands of years. The amount of such wastes will increase rapidly. We have the problem of finding a way to contain these wastes safely. Containment vessels tend to deteriorate because of the heat and radioactivity of their contents. Attempts are under way now to find a way to solidify wastes and to store them in underground vaults such as salt mines. Salt is relatively impervious to the flow of liquids. More exotic waste disposal schemes suggest sending wastes into the sun in rockets. A final philosophical question is this: Does one generation have the right to create waste problems for dozens or hundreds of later generations to manage?

In the end we are left with the basic question of whether the dangers of atomic energy justify the good of additional energy. We might hope to answer the question by turning to the Atomic Energy Commission, which is the regulatory agency for nuclear power. Unfortunately, for most of its life the AEC played the role of promoter as well as regulator. As such its credibility as regulator was sometimes questioned. For example, in 1957 the AEC issued an analysis on the results of a major hypothetical accident.(4) The study suggested that in the very unlikely event of a major accident,

---

*For some fascinating sidelights on the problem of possible diversion (theft) of fissile material—and the probable nuclear-bomb blackmail that would follow—we recommend the three-part "Profile" on Theodore B. Taylor by John McPhee which ran December 3, 10, and 17, 1973 in THE NEW YORKER.—Ed.

deaths could occur up to 15 miles, and injuries up to 45 miles. About 3,400 people could be killed and 43,000 injured. Property damage could be as high as seven billion dollars. The report caused such a strong response that in *sixteen years* the AEC has never again attempted to estimate the results of a major accident.

We must ask whether reports such as the above do more good by helping to create an informed public, or more damage by alarming the public, *perhaps* unnecessarily. We contend that there has been too much secrecy. An educated public must make its own decision based on the best data available. The government has the duty to supply that data openly and without bias.

FIGURE 6–10. San Onofre nuclear power plant. The San Onofre generating station provides 450,000 KW of electricity for Southern California Edison Company and San Diego Gas and Electric Company—enough power to supply a city of well over half a million population. Nuclear components are housed in the steel containment sphere near the center of the picture. The Pacific Ocean, in the background, is the source for water for once–through cooling. Note the major excavation almost 70 feet down from the 100-foot bluff. Lowering the plant decreases the pumping head for ocean cooling water. (*Courtesy Southern California Edison Co.*)

Looking to the future, what may we anticipate? In 1972 the AEC disavowed its promotional role and informed the operating companies and suppliers that they must do their own promoting. Late in the same year the AEC issued a first draft of a comprehensive new report on nuclear safety.(5) Critics have not found the hoped–for candor that will be necessary if confidence is to be restored in the Commission.(6, 7)

No one can say today how safe nuclear reactors are. But we need to know the answer as accurately as possible. The stakes are too great. A catastrophic accident would have at least two serious effects. First, it would result in many deaths and much property loss. Second, it would strike a major blow at the use and development of atomic power in this country. The effect on a growing "energy crisis" would be another catastrophe itself.

6.8   THE ECONOMICS OF ATOMIC POWER

The cost of electric energy generated by nuclear power plants is, as with all power plants, the sum of fixed charges and operation and maintenance charges. These costs are very difficult to estimate for design of present–day nuclear plants, and even more difficult to predict for the plants of the future. Cost estimates are much more difficult for nuclear plants than for hydroelectric or fossil–fuel plants for a number of reasons. For instance, it is difficult to estimate the cost of building the plant. We have relatively few examples to give us experience, and some early plants were partially subsidized by the government so that their cost and construction time are difficult to estimate. Furthermore, some companies underbid costs of early plants to gain entrance into the business. Also, there has been a great deal of inflation, particularly of labor costs. Finally, rapid changes in technology have occurred, along with changes in environmental or safety standards or requirements.

Fuel costs are perhaps somewhat less troublesome to predict, but the estimation process is highly complex because the spent fuel has a recovery value. Some of it is directly useful as fuel; other elements have other immediate uses, while still others can be stored for future use.

The low point in nuclear energy costs occurred in 1966, when TVA decided to build a plant at Brown's Ferry at a cost of $116 per KW, with an energy cost of 2.39 mills per KWH.

Since that time costs have been rising steadily. Present plant construction costs fall in the range of about $250 to $400/KW. The competitive position of nuclear energy vis–a–vis other forms (primarily fossil–fuel in 1972) may be declining.

In Table 6–1 we show a calculation of fuel costs which indicates the relative cost levels.(8) We now use this figure in an example of total cost of energy for a hypothetical plant.

TABLE 6–1.

Nuclear Fuel Costs (First Core Fueling)

	Mills/KWH
Raw Fuel (U_3O_8) at $7/lb.	0.40
Conversion and Enrichment	0.48
Fabrication	0.55
Reprocessing and Shipment	0.25
Gross Direct Fuel Costs	1.68
Less: Spent Uranium Credit	–.15
Plutonium Credit	–.28
Total Credits	–.43
Net Direct Fuel Costs	1.25
Fixed Charges and Interest	.40
Total Fuel Cost	1.65

In Table 6–2 we calculate the total energy for a 1,000 MW plant, assuming an 80 percent plant factor. The final result of 5.96 mills/KWH is probably an acceptable (competitive) price in many parts of the United States at this time. Again, the reader is cautioned that nuclear energy costs are highly variable at present, and the examples given here should be considered only as expressing approximate relationships among the various cost factors.

TABLE 6-2.

Total Nuclear Energy Costs

Power = 1,000 MWe   **Plant Factor = 80%**

Fixed Charges	Capital Cost ($1,000)	Cost of Money (%)	Annual Cost ($1,000)	Unit Cost (mills/KWH)
1. Depreciating capital —power plant, etc.	$200,000	13	$26,000	3.71
2. Nondepreciating capital				
a) Land	500	11.5	580	0.01
b) Operation and maintenance	1,000	11.5	115	0.02
c) Fuel cycle operations	10,000	11.5	1,150	0.16
3. Liability insurance			500	0.07
Annual fixed charges			27,823	3.97
Operating Costs				
1. Operation and maintenance			2,400	0.34
2. Fuel cost including capital cost			11,530	1.65
Operating costs			13,930	1.99
TOTAL ENERGY COST			41,753	5.96

REFERENCES FOR CHAPTER 6

1. U.S. Atomic Energy Commission, *The Nuclear Industry,* U.S. Government Printing Office, 1970.
2. A.R. Tamplin and J.W. Gofman, *Population Control Through Nuclear Pollution*, Nelson–Hall, Co., Chicago, 1970.
3. D.F. Ford and H.W. Kendall, "Nuclear Safety," *Environment*, Vol. 14, No. 7, September, 1972.
4. "Theoretical Possibilities and Consequences of Major Accidents in Large Nuclear Power Plants, WASH-740," USAEC, March, 1957.
5. "The Safety of Nuclear Power Reactors and Related Facilities, WASH-1250 (Draft)," USAEC, December, 1972.
6. R. Gillette, "Nuclear Safety: AEC Report Makes the Best of It," *Science*, Vol. 179, January 26, 1973, pp. 360–363.
7. D.E. Abramson, "The Nuclear Fast Breeder," *Environment*, Vol. 15, No. 2, March, 1973, pp. 3–4.
8. Federal Power Commission, *The National Power Survey*, U.S. Government Printing Office, 1970.

ADDITIONAL READING FOR CHAPTER 6

1. G. Bryerton, *Nuclear Dilemma*, Ballantine Books, Inc., New York, 1970.

   This is an interesting and reasonably balanced summary of the promise and problems of nuclear power. It centers on a particular proposed plant in Oregon. Besides describing nuclear reactors in laymen's terms, it also discusses in some detail the political actions which were taken in Oregon by opponents of the plant. The result is interesting, informative, and highly readable.

2. D. Nelkin, *Nuclear Power and Its Critics—The Cayoga Lake Controversy*, Cornell University Press, Ithaca, N.Y., 1971.

   This book also discusses a particular controversy, this time in New York State. There is ample political discussion, and in addition this book is concerned to a significant degree with the heat addition (or thermal pollution) problem.

3. J. Stokeley, *The New World of the Atom*, Ives Washburn, Inc., New York, 1970.

   This is a comprehensive review of the use of atomic energy in a wide variety of ways. Although it is written about the technical aspects of the problem, it is quite readable. It is suited to the non-scientific reader who wants to know more about atomic energy.

4. R. Curtis and E. Hogan, *Perils of the Peaceful Atom*, Ballantine Books, New York, 1970.

   This is one of a large number of strongly written books with the thesis that nuclear power is perhaps too dangerous to use. After presenting its view of the truth, it closes by asking the question: "Knowing this truth, do we dare continue gambling against Fate?"

5. A.W. Kramer, *Understanding the Nuclear Reactor*, Technical Publishing Co., Barrington, Illinois, 1970.

   This is a fairly short, somewhat technical book, which expands on the material presented here in Sections 6.1 through 6.6.

6. Pederson *et al.*, *Applied Nuclear Power Engineering*, Cahners Books, Boston, Mass., 1972.

This is a technical book requiring an engineering or physics background. However, it is written at an introductory level and it is not hard to follow.

7. S. Glasstone and A. Sesonske, *Nuclear Reactor Engineering*, Van Nostrand Reinhold Co., New York, 1967.

This is a comprehensive graduate engineering text in nuclear engineering.

8. *The United States Atomic Energy Commission—What It Is, What It Does*, USAEC Division of Technical Information Extension, Oak Ridge, Tennessee, 1967.

This short public information booklet describes the mission and the activities of the AEC.

9. *Potential Nuclear Power Growth Patterns,* WASH-1098, USAEC, U.S. Government Printing Office, December, 1970.

This presents the AEC view of the future of nuclear power. It involves discussion of fuel supplies, power plant forms and costs, uncertainty of predictions, and a number of possible combinations of forms of nuclear plants in the future.

10. R.V. Moore, *Nuclear Power*, Cambridge University Press, 1971.

This partly technical book reviews nuclear technology but also discusses nuclear power from another country's viewpoint. It is particularly interesting for this latter feature.

11. D.R. Inglis, *Nuclear Energy—Its Physics and Its Social Challenge* Addison-Wesley Publishing Co., Reading, Mass., 1973.

This is a good general survey of the use of nuclear energy in electric energy generation, atomic bombs, and medical purposes. Armament control, use of bombs and nuclear testing are discussed.

12. R.S. Lewis, *The Nuclear-Power Rebellion*, The Viking Press, New York, 1972.

The author wrote this non-technical book when he was also Editor of the *Bulletin of Atomic Scientists*. In dramatic style he outlines some of the severe problems of nuclear power generation and shows how alert citizens have been able to challenge long-range expansion plans which could be or which are dangerous to the biosphere. For example, Chapter 6, "Boot Hill," describes in urgent terms what can happen when radioactive wastes leak, as they are now doing in Richland, Washington, and what can happen when bureaucracy decides to appropriate specific tracts of land for further waste storage.

PROBLEMS FOR CHAPTER 6

General Problems

6.1. How many pounds of U^{235} are there in one ton of Uranium? (14 lbs.)

6.2. Explain what is meant by $_{94}Pu^{239}$. Indicate all of the information given.

6.2. What is the difference between fissile and fertile isotopes?

6.4. Make a chart of the three forms of radiation discussed in this chapter, comparing the structure, relative weights, ability to penetrate matter, and biological effects.

6.5. Explain the purpose of control rods, moderators, coolant, and containment vessels.

6.6. Explain the meaning of the term LMFBR, indicating the significance or meaning of each letter.

6.7. How many pounds of mine waste are produced to obtain one pound of U_3O_8?

6.8. What percentage of all the energy produced in 1964 was used by nuclear processing plants?

6.9. List the major advantages and disadvantages of nuclear power.

6.10. If a radioactive material has a half life of 10 years, what fraction of the material remains at the end of 40 years?

Advanced Mathematical Problems

6.11. What is the approximate amount of mass lost by $_{92}U^{236}$ when it fissions?

6.12. How many half-lives are required for 99.9% of a radioactive material to decay?

6.13. Draw a logarithmic chart of the effect of various levels of radiation in mrems. Plot the mrems vertically, and indicate significant levels such as background radiation, AEC limitations, fatal exposure levels, common power plant levels, etc.

6.14. Assume that the percentage of power generated by atomic plants rises exponentially from 1 percent in 1970 to 50 percent in 2000. a) Find the percentage of power generated by atomic power of 1980 and 1990. b) Plot the percentage of power generated by atomic power from the year 1970 to 2000. Use semi-log plotting paper. c) What is the doubling time for percentage growth?

6.15. For the example given in Section 6.8, find the energy cost in mills/KWH if raw fuel costs increase from $7/lb. to a) $10, b) $50, c) $200.

6.16. For the example given in Section 6.8, find the maximum acceptable cost of raw fuel if it is competing with fossil-fuel available at 7.5 mills/KWH.

Advanced Study Problems

6.17. Write a short paper indicating whether or not you favor the further growth of atomic power. Give your reasons.

6.18. Write a short paper on the history and future development plans of the breeder reactor. Include a forecasted timetable of development milestones such as date of first active large-scale plant, date when breeder plants could produce a significant amount of power (say 10 percent of U.S. capacity), etc.

6.19. Read at least two articles or books on reactor safety written by proponents (see, for example, power industry journals available in most libraries), and two written by opponents (environmental magazines, or Reading 4 or Reference 2 above). Compare specific points of disagreement, and decide whether you can make a decision about safety based on the evidence given. You will have to evaluate this evidence critically.

6.20. Study and report on the history and critical reaction to the AEC document WASH-1250. (See References 5, 6, 7 and more recent developments which should become available.)

6.21. Read an Environmental Impact Report for a nuclear power plant. Summarize the most important factors of the report. Discuss how the report will be used by the government.

7

GEOTHERMAL POWER

That I may tell pale-hearted fear it lies,
And sleep in spite of thunder.

Macbeth

The interior of the earth is a mass of very hot and sometimes molten rock. The heat from this region is a huge potential source of natural energy which can be used for the generation of electric power at many places in the world. In this chapter we talk about where geothermal development is possible and why. We then center our attention on the major development presently underway in the United States.

The basic principle of geothermal generation is that steam is used to drive a turbine, as in thermal plants fueled by Uranium or by fossil fuels. The essential difference in geothermal power is that the source of heat which produces the steam is the earth's interior. This heat is tapped by wells drilled as much as two miles into the earth.

7.1 PANGAEA AND TECTONIC PLATES

One of the most fascinating aspects of geothermal power is that it is closely linked to a newly–evolving and extraordinary hypothesis about the earth's geological history. Scientists now believe that the world once had only one land mass called Pangaea (all lands), and that the present continental land masses resulted from the breakup of Pangaea and the movement of certain earth masses over the past 200 million years.(1) South America broke away from Africa, North America from Europe, and in perhaps the most spectacular dash of all, India raced from Antarctica at a speed of perhaps inches per year, until it smashed into Southern Asia, creating the Himalayas.

The theory holds that the earth has an outer shell or crust (or *lithosphere*) about 50 miles thick. This is broken into perhaps ten major pieces called *tectonic plates*. For reasons which are not clear these plates move with respect to each other, thereby rearranging the earth's land masses. The boundary lines between plates, which are thus cracks in the earth's crust, are called "rifts." Along these rifts tearing or scratching takes place as one plate moves past the other. When such tearing occurs it is felt as an earthquake on the surface of the earth. One such famous rift area is near the west coast of the United States. The San Andreas Fault, whose movement caused the disastrous 1906 quake in San Francisco, is associated with this rift. (According to present theory Los Angeles should drift along the west coast rift area as far north as San Francisco in about 10,000,000 years.) Other major rifts are located near other parts of the world which typically experience severe earthquakes. These include Japan, western South America, Turkey, and many others.

If the moving plate theory and some of its implications are true, then it is no coincidence that most of the earth's geothermal power sites are located near rifts. Or, to tie in the factor above, geothermal sites are often located in major earthquake zones. Volcanic activity is also common in such areas.

There is an explanation in the plate theory justifying geothermal sites near rifts. It is theorized that at cracks or rifts in the earth's crust, hot molten material can seep or well up from the earth's interior to within a couple of miles or so of the earth's surface. Here the hot material may encounter underground water, resulting in steam or hot water. Sometimes this steam or hot water reaches the earth's surface naturally, appearing as geysers or hot springs. At geothermal sites, wells are drilled to reach the best steam or water available.

7.2 GEOTHERMAL ENERGY

Geothermal heat almost certainly originates with magma (molten rock) which is relatively near the earth's surface, probably for the reasons suggested in the previous section. Figure 7–1 suggests how the earth's interior heat may be converted to steam or hot water. The lower layer is the hot magma. It is covered by an impermeable crystalline rock perhaps one to two miles thick, which transfers heat up to a layer of porous rock. This layer is topped by a layer of less–permeable matter. Water finds its way down to the porous rock, and because of the heat transferred up from the magma, the porous rock region acts almost like a boiler in changing the water into hot water, wet steam, or dry steam. Sometimes this water or steam escapes to the earth's surface appearing as fumaroles (emissions of hot gases or steam) or geysers (emissions of a mixture of steam and hot water).

Geothermal steam or hot water is obtained by drilling down into the porous rock "boiler." The best possible product of a geothermal well is steam that is hot, dry, and clean. It should be hot to make the thermal plant efficiency high. It should be dry—containing few water particles—to minimize the heat loss to the water and the corrosion resulting from wet steam. Finally, it should be clean; that is, it should contain as few mineral impurities as possible, because impurities cause turbine corrosion and leave undesired mineral deposits on the earth's surface.

The quality of the medium varies greatly from one site to another. The best sites produce fairly hot dry steam; the poorest produce hot water with a high concentration of minerals. The like-

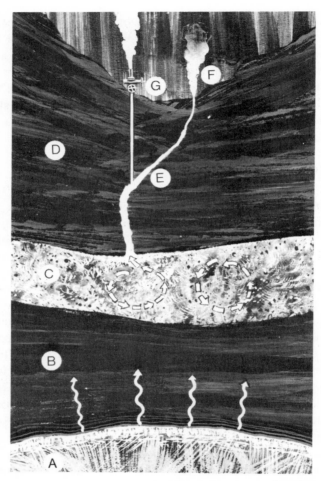

FIGURE 7—1. Cross–section of geothermal region. In certain places through-out the world, nature provides just the right combination of circumstances to produce steam which can be used for geothermal power production. Hot magma (A) wells up to within a few miles of the earth's surface, near crustal fracture lines. Heat from the magma passes through a region of low–permea-bility (that is, nearly solid) rock (B) up to a zone of porous rock (C). Region C is nature's boiler. Here, water that has seeped down, usually over decades, is transformed into steam. A low permeability layer (D) caps the boiler to hold the steam. Thin breaks (E) in the cap let steam escape and appear at the surface as geysers or fumaroles. Then man drills a well (G) to capture the steam for use in a geothermal plant. (*Courtesy of Pacific Gas & Electric Co.*)

lihood that a site can produce economically feasible geothermal power is closely related to the quality of the medium.

At some sites the steam or water, pumped up from one to two miles down, is as much as 50 to 100 years old. That is, it has taken that amount of time for surface water to seep down through the low–permeability top layer into the steam–producing porous-rock region. This suggests that steam beds may be exhausted in time since replacement water is so slow in reaching the underground "boiler." Artificial insertion of water through old wells or dry wells may or may not turn out to be practical. At the only site where geothermal plants have been operating for many years, wells have dried up in 20 to 30 years. There is not sufficient experience or thorough understanding of the mechanism to predict how long new wells may produce steam. When a well does dry up, it is often possible to drill a new well in the same general vicinity. In other cases a particular steam field may become exhausted. This uncertainty factor is important in an industry which is responsible for meeting the power demand of its customers reliably.

7.3 GEOTHERMAL POWER AROUND THE WORLD

The rifts or breaks in the earth's 30 mile deep crust are widely spread over the earth's surface. Hence it is not surprising that geothermal power has been developed or is being considered in many countries around the world.(2) Ten countries have developed or are constructing geothermal plants. An additional 14 or more countries are also seriously considering this form of power. The ten active countries are shown in Table 7-1 below. (Adapted from Reference 2.)

Use of geothermal heat for space heating certainly predates man's earliest histories. The first important generation of geothermal *electric* power came in Larderello, Italy in 1904. At first this source lit only a few light bulbs. In time however, it became the major source of power for Italy's electric railroads. Today (1973) it is still the largest operating geothermal source in the world. The Geysers system north of San Francisco in the United

TABLE 7-1.

Active Geothermal Production

Country	1969 MW	Planned MW	Total MW
El Salvador	0	20	20
Iceland	0	3	3
Italy	384	0	384
Japan	32	237	259
Mexico	3	150	153
New Zealand	170	120	290
Taiwan, China	0	10	10
Turkey	0	30	30
USSR	5	26	31
United States	83	400	483

States has moved into second place in size, and if present plans are carried out, will become the largest producer of geothermal power by 1973 or 1974. These two fields, plus one at Matsukawa in Japan, are the only three active sites presently producing dry (slightly superheated) steam.

Other nations are developing plants with lower-quality steam. Such plants tend to be more costly than dry steam plants. It is impossible to give a comprehensive picture of world geothermal activity because new fields are constantly being explored and ways to utilize low quality heat are being sought. The prospects for world-wide development are truly extraordinary, but it is too early to say where these prospects lead. Rather than try to summarize the world picture, we turn now to a discussion of the United States, where a highly successful geothermal power plant complex has been developed in a dry steam field, and where solutions are being sought in a very wet and very contaminated field.

7.4 GEOTHERMAL POWER IN THE UNITED STATES

Development and exploration of geothermal power in the United States has been largely limited to the State of California. Some important exceptions include the discovery of some dry steam in Yellowstone National Park (Wyoming), development of a

10 MW plant near Brady, Nevada, and the use of geothermal heat in about 400 buildings in Klamath Falls, Oregon. A large number of other sites, particularly in the western United States, no doubt justify and will experience exploration.

The major success in the United States is The Geysers in northern California, which we describe in detail in the next section.(3)

A plan under study in the United States calls for exploding an atomic bomb deep beneath the surface of the earth and using the trapped heat to generate steam.

Without a doubt the greatest interest in an untapped potential geothermal source is in the Imperial Valley—Salton Sea area of southern California. Estimates have been made that this region might yield as much as 20,000 to 30,000 MW of electric power, nearly equal to the present *total* power capacity in the State of California. Study of this potential is being spurred by the successful development of the Cerro Prieto plant (75 MW) just 20 miles or so south of the border in Mexico. Cerro Prieto steam is relatively wet and contaminated by minerals. To the north in the United States the geothermal medium tends to be much more contaminated in wells drilled to date. Such wells, in this region, offer three opportunities for commercial development. The first is geothermal electric power. The second is the production of minerals from the medium, including Potash, Lithia, Salt, Calcium Chloride, Manganese, Copper, Lead, and Silver. The third is the production of desalinated water in a very dry region. Obstacles to development include the corrosive effect of the minerals and the necessity of getting rid of the residue, probably by pumping it back down a dry well.

The extraordinary heat reserves in southern California have recently led to great optimism about the prospects for this region. Evidence of heat concentrations in the area south of the Salton Sea comes from infra-red satellite photographs, surface temperature tests, and some preliminary drilling. A Bureau of Reclamation document describes a very ambitious plan for research and development over a seven-year period.(4) This would be followed by a demonstration program which would combine power generation and water desalination. One hundred twenty five thousand acre-feet of water would be pumped from the Salton Sea; 100,000 acre-feet of desalinated water would be returned to the Salton Sea to control the salinity of that body. Twenty five thousand acre-feet would be used in cooling towers (see Chapter 12) for 420 MW of geothermal power. Power cost would be a highly attractive five mills per KWH. The final stage of the project would deliver 2.5 million acre-feet of desalted water annually to the Colorado River and would produce 10,500 MW of power.

A recent *Wall Street Journal* article indicates an optimism for development in this area suggested by major land leases and well drillings by large energy corporations.(5)

Will this great potential resource be developed? The answer hinges principally on whether the technological problems can be solved at a cost which is competitive with alternative generation schemes. It will certainly be most interesting to watch this area.

7.5 THE GEYSERS

The Geysers area, located about 75 miles north of San Francisco, California, is the fastest growing geothermal power producing region in the world. In 1973 it had a capacity of 302 MW. Present plans are to add 110 MW per year in 1973 and 1974. Future development could lead to a total capacity of 1,000 MW by 1980. Estimates of the final capacity of the field range from 2,000 to 4,800 MW.

All of the power presently generated at this site belongs to the Pacific Gas and Electric (PG&E) Company. The present capacity represents a little less than three percent of the company's present total generation capacity. This figure will increase by about one percent of total capacity per year for at least the next two years.

Units built in recent years have been 55 MW plants. Each such plant receives steam from about 10 wells. In 1974 a single 110 MW plant will be built. In general, geothermal plants tend to be smaller than other thermal plants because of the lengths of steam pipes required to feed a large plant. Wells must be separated by some minimum distance to avoid overlap of steam demand. Between 20 and 40 acres are typically dedicated to each well. If a very large plant were built it would be necessary to pipe steam in from greater distances with a corresponding loss in heat and increase in pipe costs.

Wells at The Geysers are drilled to depths ranging from 3,000 to 9,000 feet. The bottom of the producing field has not yet been reached. Also, it is not known whether or not there is a single large reservoir about seven miles in length by two miles in width or a

FIGURE 7–2. The Geysers geothermal site in Northern California. Far below the surface of the earth, the immense heat of the earth's core creates great pockets of steam which can be used to drive turbines. Steam at the Geysers site is hot and clean, making it a very attractive source of energy. The life of such steam pockets is very difficult to estimate. Other sites are under exploration in the United States and around the world. (*Courtesy Pacific Gas and Electric Co.*)

series of smaller pockets of steam. Steam wells produce from 50,000 to 150,000 lbs./hr. of dry steam. Gaseous impurities average about one percent of the steam flow. The most noticeable gas released is hydrogen sulfide, which causes a slight odor of rotten eggs in some parts of the region. The steam also contains a small amount of rock dust.

It is believed that the wells at The Geysers can and will be depleted. Accordingly, capital investment is being amortized over the expected (or guessed-at) life of the reservoir.

Steam at The Geysers is supplied to PG&E by three independent companies: Union Oil, Thermal Power, and Magma Power. They make geological studies, drill wells, build roads, build steam lines to the plants, and return certain waste products back into the ground via dry wells, as will be discussed later.

We see how PG&E uses the steam obtained from its suppliers through a discussion of a typical Geysers geothermal power plant, sketched in Figure 7-3.

Hot dry steam comes in at the upper left from suppliers' pipelines. It is first passed through centrifugal separators which remove any rock particles or dust from the steam. (At other sites around the world such separators may be used at this point to separate steam from water or water vapor, if the medium is fairly clean and dry. At more contaminated sites the geothermal medium is not used directly in the turbine but is used to produce steam in a closed loop through some kind of heat exchanger.)

The hot steam next passes through a rather low pressure steam turbine. The pressures available here are not comparable to the much greater pressures generated in conventional modern thermal plants. The expanded steam passes out of the turbine. At The Geysers it goes to a condenser. A reasonable question is why we do not simply vent the used steam into the air, since there is no problem here of recycling the steam-water as in most thermal plants? There are two answers. The first is that the expended steam contains impurities which are undesirable in the environment. The second reason is that the condenser works at a *back pressure* (4″ Hg. compared to atmospheric pressure near 30″ Hg.). This back pressure is like a high quality, though not perfect,

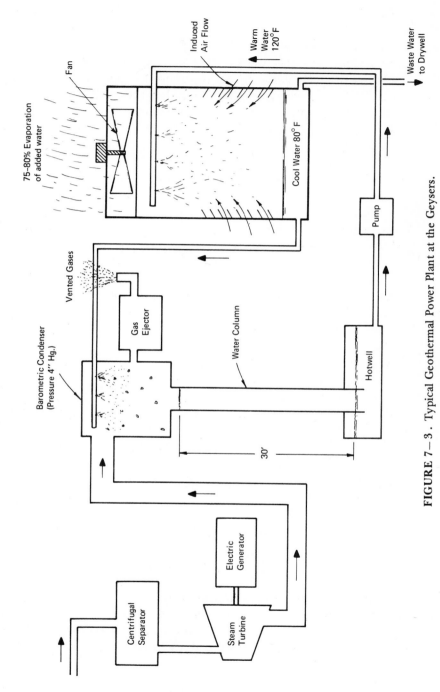

FIGURE 7–3. Typical Geothermal Power Plant at the Geysers.

vacuum which helps "suck the steam out of the turbine." This has the effect of increasing the efficiency of the plant. Even with the condenser, efficiency is low (around 15 percent) because steam pressure and temperature are low. Thus relatively more steam must be used here to generate a given amount of power, compared to a modern thermal plant. But the steam is inexpensive and no boiler or fuel is needed, so the cost turns out to be competitive with that of conventional plants.

The condenser used is a so-called "barometric condenser." Steam flows into the condenser, where it encounters a spray of cool water which condenses it by contact. This is the same basic principle used in Newcomen's engine (see Figure 5-1b). When the steam condenses, it leaves behind a high quality vacuum. The effect is that the condenser acts much like a huge water barometer, with the water column in the barometer's pipe standing about 30' above its "hotwell" pool, which is at atmospheric pressure.

Example 7-1

On the earth's surface atmospheric pressure is about 30" of mercury (Hg.). That is, a column of mercury in a cylinder with an evacuated and sealed top and a bottom immersed in mercury which is exposed to the atmosphere will rise to a level of about 30" at sea level. Find the height of water in a water barometer at sea level.

The specific gravity (in comparison with water) of mercury is 13.546. Since water is this factor lighter than mercury, we would expect it to rise by that factor more than mercury.

$$30'' \times 13.546 = 407'' = 34'$$

Since the back pressure in the condenser above is 4" we can expect a water rise of:

$$26'' \times 13.546 = 353'' \cong 30'$$

The steam from the ground contains a certain amount (about one percent) of gases which do not condense. If these were allowed to accumulate in the condenser they would gradually destroy the vacuum. A three-stage, steam-driven, gas ejector is used to remove these gases. The ejector requires about five percent of the steam used in the plant.

From the hotwell at the bottom of the condenser, water is pumped to the top of a forced–draft wet cooling tower. (See also Section 12.3.) The water is sprayed down (at 120° F) over baffles which slow its descent, as air is pulled in from the sides by a fan (which looks much like a propeller) at the top of the tower. By the time the water reaches the bottom of the cooling tower, which is 50–75′ in height, its temperature has been reduced to about 80° F. About 75–80 percent of the water circulated through the tower is evaporated into the atmosphere. Of the remaining cooled water a part is pumped back to the barometric condenser to use as the spray water in the condensation part of the cycle. About 20 percent of the water, containing most of the undesired contaminants, is returned to the ground through dry wells.

7.6 ENVIRONMENTAL FACTORS

From an environment standpoint, geothermal power is generally considered to be quite "clean," or desirable. It is necessary, however, to take certain precautions to protect the environment, depending on the local conditions.

One advantage of geothermal power is that it does not use up irreplaceable fuels, unless we consider the heat lost from the earth's interior. This heat, however, is so huge that there is no way that significant amounts could be taken from the earth by geothermal power even if it were developed at many times its present level for hundreds or thousands of years.

A second advantage is that relatively small areas of land are used compared to land use in the other "free-fuel" systems, namely hydro– or solar power. Large fields with well-heads, steam pipes, power plants, and cooling towers might cover a few square miles. There is sometimes excessive noise due to escaping steam, and minor odors from escaping gases. These factors as well as aesthetic concerns may make geothermal plants undesirable as near neighbors, but the effects should not extend to very great distances.

FIGURE 7—4. Geothermal exploration. Engineers drill for geothermal steam near Mono Lake in Eastern California. It has been estimated that major parts of the western United States may yield geothermal energy. But first we must find the steam, and then harness it. Whether geothermal power will make major contributions to the nation's energy needs or not is one of the questions for the rest of this century. (*Courtesy Southern California Edison Co.*)

A major potential problem concerns the possible destructive effects of contaminated waste water. Fortunately, it is usually possible to dispose of such wastes, when necessary, in an acceptable way. As an example, early in the history of The Geysers, waste waters were released to a nearby creek. It was found that these wastes contained Ammonia, which is harmful to fish, and Boron, which is harmful to plants. Today all waste water at The Geysers is returned to the earth through non-productive wells. Other areas must deal with wastes in similar or alternative ways, as appropriate.

Another possible problem is *land subsidence* (sagging or falling of land) resulting from the removal of large amounts of steam or water from the ground. Relatively little evidence is available as yet on the likelihood of subsidence. Some subsidence has occurred at oil field sites, and at geothermal sites in Mexico and New Zealand. One approach to the problem is to re-inject water into the the ground.

Similarly, little is known about possible seismic effects. Earthquakes could be triggered by geothermal activity such as water re-insertion. Conceivably, however, the opposite effect may occur. Water insertion in seismic areas has often been suggested as a way to reduce the danger of major earthquakes by inducing minor quakes. The water acts as a lubricant to help the tectonic plates slide.

Geothermal plants will release a large amount of waste heat to the environment. Whether or not this release is environmentally serious is difficult to say.

So it appears that there are some important questions about the environmental effects of geothermal power production. On balance, however, this approach to generation is perhaps as attractive as, or more so than, any present alternative.

7.7 GEOTHERMAL POWER ECONOMICS

It is difficult to summarize the economic status of geothermal power because costs are highly dependent on the country where the site is located, and on the quality of the steam or water available. In determining whether geothermal power is economically desirable, it is of course also necessary to consider the costs of alternative generation schemes. These costs vary greatly from one region to another. In countries where hydro power is plentiful, geothermal power may not be competitive. However, we can say fairly generally that geothermal power, using good quality steam, is quite likely to cost less than any conventional thermal process, atomic or fossil. Geothermal costs may range from 60 to 90 percent of conventional thermal costs.

At Larderello, Italy the KWH cost is about 3.2 mills.(2) At Matsukawa, Japan costs range from 4.6 mills/KWH for a 20 MW plant down to an expected 3.1 mills/KWH for a second plant. At The Geysers in the United States costs have varied from 4.5 to 5.0 mills/KWH in the initial plants, and are expected to decrease somewhat as plant size increases. Fossil–fuel electric energy costs in California tend to vary from about six to nine mills/KWH.

The major advantage of geothermal plants is in the lower capital costs (fixed charges) of the plant itself, because the boiler has been eliminated. Fuel costs tend to be roughly comparable in fossil–fuel, nuclear and geothermal plants. In the latter the fuel costs relate to the cost of exploring and drilling for steam and delivering it to the power plant.

Steam can be obtained by the utility company producing the power, or it may be purchased by them from independent companies. The latter arrangement may be convenient or perhaps even necessary if the steam fields are not owned by the utility, or if regulations prohibit the utility from speculative explorations. At The Geysers, the steam is owned by three independent companies and sold to PG&E for 2.7 mills per net KWH of energy generated at the site. This cost includes the cost of returning waste material to the earth.

For its newest units, PG&E has estimated total fixed charges per year of $1,851,000 for two 55 MW plants. These plants have a capacity of 55 MW and operate at 80–90 percent plant factor. Assuming a plant factor of 85 percent, we can calculate the fixed charges per energy unit as:

$$\frac{\$1,851,000}{110,000 \times 8,760 \times 0.85} = \$2.26 \times 10^{-3}/\text{KWH} = 2.26 \frac{\text{mills}}{\text{KWH}}$$

Hence total energy costs are:

1. Steam cost (from supplier) $2.70 \dfrac{\text{mills}}{\text{KWH}}$

2. Fixed charges, including plant
   operating costs (85 percent P.F.) $\underline{2.26}$

$4.96 \dfrac{\text{mills}}{\text{KWH}}$

This cost figure is quite attractive in the PG&E service area at this time.

REFERENCES FOR CHAPTER 7

1. R.S. Dietz and J.C. Holden, "The Breakup of Pangaea," *Scientific American*, October, 1970, pp. 30–41.
2. G. Facca, "General Report on the Status of World Geothermal Development," *United Nations Symposium on the Development and Utilization of Geothermal Resources*, Pisa, Italy, 1970.
3. D.B. Barton, "The Geysers Power Plant—A Dry Steam Geothermal Facility," Geothermal Resources Council Meeting, El Centro, California, February 16–18, 1972.
4. "Geothermal Resources Investigations—Imperial Valley, California," Bureau of Reclamation, January, 1972.
5. E.C. Gottschalk, Jr., "Steam Below Ground Seen Giving Big Boost to U.S. Energy Supplies," *Wall Street Journal*, March 20, 1975.

ADDITIONAL READING FOR CHAPTER 7

1. *Continents Adrift*, W.H. Freeman and Co., San Francisco, 1973.
   This is a very interesting collection of readings from *Scientific American* on continental drift and its relation to various phenomena such as earthquakes and the earth's heat. Geothermal power is not discussed directly.
2. D. Tarling and M. Tarling, *Continental Drift*, Anchor Books, Doubleday and Co., Garden City, N.Y., 1971.

This book also discusses geologic factors relating to geothermal power, but does not consider such power directly.

3. M. Goldsmith, "Geothermal Resources in California—Potentials and Problems," *Environmental Quality Laboratory Report No. 5*, California Institute of Technology, December, 1971.

This is an excellent general review of the geothermal potential in California. It is written at a semi–technical level.

4. "The Economic Potential of Geothermal Resources in California," Geothermal Resources Board, State of California, January, 1971.

This report stresses economic factors relating to geothermal development. It also reviews briefly some of the potential areas of future growth.

5. J.R. McNitt, "Exploration and Development of Geothermal Power in California," *Special Report 75*, California Division of Mines and Geology, 1965.

This document is not as up–to–date as 3 and 4, but it does have an interesting historical and geological review of California geothermal power development. It also includes some technical material on geothermal power plant configurations.

PROBLEMS FOR CHAPTER 7

General Problems

7.1. Consider Figure 7–3. If 75–80 percent of the condensed steam ends up being evaporated and vented by the cooling tower, what advantages are there over simply venting the steam out of the turbine?

7.2. List the possible environmental problems related to geothermal power and indicate solutions where such exist.

7.3. At The Geysers power plant, forced–draft cooling towers are driven by 150 horsepower fans (or propellers). Five such towers are required by each 55 MW plant. Determine the approximate percentage of the 55 MW generated that must be used to drive these fans. (About one percent)

7.4. What factors might tend to increase the costs of geothermal power in areas where the medium is hot, dirty water rather than dry steam?

Advanced Mathematical Problems

7.5. What would be the height of a column of methyl alcohol in a barometer at sea–level pressure (say about 30″ Hg.)?
7.6. For the example given at the end of the chapter, plot the cost of energy versus plant factor for a plant factor ranging from 50 to 90 percent.

Advanced Study Problems

7.7. Discuss how geology is a major feature in the generation of power by hydro, fossil–fuel, nuclear, geothermal and tidal schemes. That is, describe how geology affects each scheme.
7.8. Construct a table of locations on the earth where two or more of the following phenomena are common: earthquakes, geysers or fumaroles, volcanos, geothermal power exploration, any other correlatable phenomena. Note that you are not *proving* a necessary connection by this identification, but only pointing out where certain phenomena coincide. Such evidence is certainly not sufficient to prove any major geological hypothesis about the earth's crust though it might be used in partial support of a theory.
7.9. Describe in more detail than is given here the most recent commonly accepted concept or hypothesis about the form of the earth's interior, its source of heat, its major components, and its history.

7

8

TIDAL POWER

On such a full sea are we now afloat,
And we must take the current when it serves,
Or lose our ventures.

Julius Caesar
Shakespeare

For centuries man has realized that the ebb and flow of the tides offer a source of energy because of the potential energy of raised tidal water or the kinetic energy of tidal streams. The purpose of this chapter is to consider this source of electric energy. To date there has been relatively little development of tidal projects despite the fact that such schemes seem so attractive. We shall ask why the apparent promise of the tides has not led to very much development. It is a common experience that many projects which seem attractive develop slowly if at all. One reason for devoting a chapter to tidal power is to see why this is so.

Man has used the tides as a source of energy for countless centuries. The Domesday Book (11th century) mentions a tidal mill at Dover, England. Salem, Massachusetts had a mill in 1635. Tides successfully drove a pump that supplied London with its water until its demolition in 1842.

In more recent years many potential sites around the world have been considered for electric power plants. Some, such as the Passamaquoddy site on the U.S.–Canada border, have been studied very extensively. One major plant, at La Rance in France, was completed in 1966.

8.1 ENERGY FROM THE TIDES

The originating source of tidal energy is in the kinetic energy of the orbiting and rotating earth, moon, and sun. As long as these bodies move relative to one another, the waters of the earth rise and fall due to changing gravitational effects.

Energy is available because water changes its level or height. This is equivalent to the development of a head in a hydroelectric project. There are at least three ways in which tidal energy might conceivably be harnessed. These ways are indicated in Figure 8–1.

The first way of harnessing tidal energy is simply to place a water wheel in a tidal stream, as shown in Figure 8–1a. This is analogous to using a water wheel in a river. The problem with this scheme, as with the water wheel in the river, lies in the variability of the tidal stream flow. A tidal wheel may be acceptable for some applications, such as pumping water or milling grain, but it is almost certainly unsatisfactory for electric energy generation.

The second possible scheme is shown in Figure 8–1b. A large floating object such as a barge is raised by an incoming tide; it is constrained between pilings. It can then be held and dropped later or allowed to fall with the tide. In either event, as it falls it drives an electric generator. To determine the practicality of such a scheme we consider a specific example.

Example 8–1

Consider a ship of 20,000 tons displacement, with a tidal rise of 15 feet. What is the average power which can be generated?

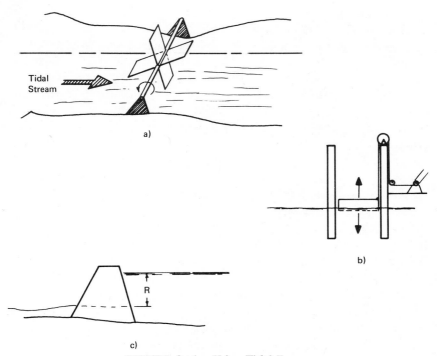

FIGURE 8–1. Using Tidal Energy.

The average power in foot-pounds per second is $P = \dfrac{HW}{T}$, as given in Equation 4–2 where H is the head or change in level in feet, W is the weight in pounds and T is the time over which the average is taken in seconds. We assume that the time between tides is about 12 hours (43,200 seconds).

$$P = \frac{15 \times 20,000 \times 2,000}{43,200} = 13,900 \text{ ft.-lbs./sec.}$$

$$= 25.3 \text{ HP}$$

$$= 18.85 \text{ KW}$$

Hence our 20,000 ton barge can generate about enough power for a few private residences. This won't do, of course. Let's stop and take a look at the situation and see why we are not getting more power. Recall our basic power equation:

$$P = \frac{HW}{T} \tag{8-1}$$

One problem here is with T. The available energy (HW) must be averaged over 12 hours. The result is that P is very small. We can't do anything about T since it is fixed by orbits of the earth and moon. The only way we can increase H is by going to some place on the earth where the difference between high and low tides is greater. 'This we can try to do. Next our attention centers on W. One approach is to use more barges. Clearly, however, if one large barge yields only 19 KW, we shall need far too many barges to obtain a reasonable level of power. Barges are not the answer.

Raising and lowering the barge is equivalent to raising and lowering the amount of water displaced by the barge. The obvious solution is to work with the moving water rather than with an object floating on the water. We can do this by building a low dam (Figure 8-1c) across the mouth of a bay or tidal estuary. As the tide comes in, gates are opened in the dam and water flows into the bay or estuary. At high tide the gates are closed. When the tide-water outside the bay recedes, a head develops which is used to turn hydraulic turbines as in a hydroelectric plant.

The weight of the water stored behind the dam, from the high tide level to the low tide level, is RSw, where R is the tidal range in feet (or difference between high and low tide levels), S is the average surface area of the storage pond or reservoir in square feet, and w is the weight density of water in pounds per cubic foot. Consider again Figure 8-1c. We would like to run this weight of water RSw through the turbine, but we cannot assume the head is R because as water flows out of the reservoir, the level of the reservoir decreases. The energy actually available can be obtained by taking the head as the average difference in water level. This average value is $\frac{1}{2}$ R. Hence the available energy is:

$$E = \frac{1}{2} R \times W = \frac{1}{2} R \times RSw$$

$$= \frac{R^2 Sw}{2} \tag{8-2}$$

The average power is obtained by dividing by T.

$$P = \frac{R^2 Sw}{2T} \qquad\qquad (8\text{-}3)$$

The value we use for T here depends on the form of the system. In some configurations both incoming and outgoing tides are utilized. In this case T can be taken as six hours (21,600 seconds).

Example 8-2

Consider a tidal reservoir with an area of 10 square miles and a tidal range of 17 feet. Assume T = 6 hours. Find the average power.

$$P = \frac{R^2 Sw}{2T} = \frac{(17)^2(10 \times 5,280^2)(62.4)}{2(21,600)}$$

$$= 115,800,000 \text{ ft.-lb./sec.}$$

$$= 209,000 \text{ HP } (159,000 \text{ KW})$$

This is a fairly respectable average power, and the example goes a long way in support of our intuitive sense that there's a lot of free energy going to waste in the tides. This is the kind of statement we often hear in regard to tidal power. Similar statements are made with respect to many other apparently attractive schemes. Why don't we have hundreds or thousands of tidal power plants? It is true that sometimes technological "progress" is constrained by a lack of imagination or the courage to try something new. But the history of electric energy generation is full of examples of great engineering challenges met in the face of great opposition and formidable technical obstacles.

The basic problem with the development of tidal projects is that the energy is *not* free, at least not as electric energy. Dams must be built in salt water basins, usually in the presence of massive amounts of rapidly flowing water. It is often very expensive to build such facilities. We cannot argue, of course, that once the facilities are built, the energy is free, because we must pay for the capital investment over the life of the project, as we saw in Section 4.9 in our discussion of the economics of hydroelectric plants. The initial cost of the plant may be so great that the cost of energy per KWH is greater than that available from alternate generation methods.

8.2 TYPES OF TIDAL POWER SYSTEMS

We have described a simple tidal power system in which water is allowed to enter the bay or estuary, either through gates or turbines, with a rising tide, and is used to generate electric energy when the tide falls. This is called a *one–basin* scheme. This scheme has the disadvantage that the power available varies with the tide conditions. Of course the power availability will not in general match the power demand in time. As we have seen we have no way to store electric energy except through a secondary approach such as the pumped–storage scheme discussed in Chapter 4.

An alternative arrangement is the so–called two–basin scheme. The basic scheme is sketched in Figure 8-2a. Two basins are separated by a dam containing the turbogenerator units. One basin, called the low pool, is permitted to empty itself with the low tide and is then sealed off to the incoming tide by means of gates. The other basin, called the *high pool*, is allowed to fill with the incoming tide. The water in the high pool flows through the turbines into the low pool. When the tide has begun to ebb, the sea gates of the high pool are closed. When the tide drops lower than the level of the filling low pool, the low pool's gates are opened to release its water. When the tide comes in the whole process is repeated. The basic objective of this scheme is to increase the versatility of the system by providing for power generation at any time. This advantage over the single–basin scheme costs money, because an additional system of dams is required.

The timing of water flow in the two–basin scheme is illustrated in Figure 8-2 (b and c). Figure 8-2b suggests two types of power load which might be placed on a system. Between times t_a and t_b a fairly constant load is presented. At t_a the upper pool begins to release water to the lower pool, generating power P_1. As the high tide level exceeds the high pool level (at t_e), the high pool is refilled. The high pool continues to release water through the turbo-generators into the low pool until t_b, when the power demand ends. The high pool level then remains constant until the next high tide, or the next power demand. In the meantime, at t_f the low tide reaches the level of the low pool, which then empties into the sea. The basic process is repeated for the short peak demand between t_c and t_d. These are two examples of demand types which can be met. The two-basin scheme could meet base loads, peak loads, or any combinations, depending of course on its total capacity. (See also Reference 1.)

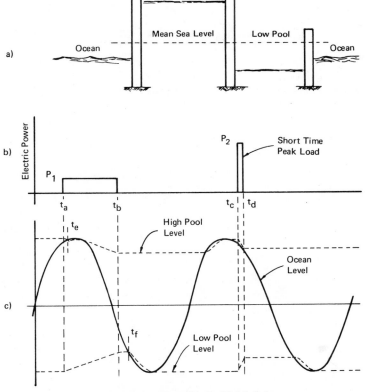

FIGURE 8-2. Two-Basin Tidal Scheme.

8.3 TIDAL POWER SITES

In seeking a potential site for a tidal power plant the following four criteria provide a first-order evaluation:

1) The tidal range R should be large.
2) The storage area S should be large.
3) The site should allow the development of the necessary plant for reasonable cost.
4) Construction of the plant and impounding of the tidewater should not have an unacceptably adverse effect on the environment of the surrounding region.

One way to evaluate a tidal site is to estimate its maximum energy, as given by Equation 8-2. For one complete tidal cycle this becomes:

$$E_{max} = R^2Sw \qquad\qquad (8-4)$$

If the energy E_{max} is averaged over the time required for a full tidal cycle (12 hours and 24.6 minutes or 4.46×10^4 seconds) we obtain what we might call P_{max}.

$$P_{max} = \frac{R^2Sw}{4.46 \times 10^4} \text{ (ft.-lbs./sec.)}$$

If we now multiply P_{max} by the number of hours in a year (8,760), we obtain the maximum possible annual energy, E_{AN}. In practice we can obtain only a fraction (perhaps 10-20 percent) of this maximum energy. The head cannot be completely utilized over a time which is any reasonable part of the tide cycle because the difference between levels decreases from the maximum. Also, the turbine itself has a size which is significant compared to the head. These factors and others conspire to limit the amount of energy we can obtain in practice.

P_{max}, E_{AN}, R and S are shown in Table 8-1 for a number of potential tidal sites around the world. This list is by no means exhaustive.

It is apparent that there is a great deal of energy available in some of the world's tidal pools. The question is, why is that energy not used? As we saw before, the answer often is that it simply is not economical to use this energy. To explore this point in more depth we will finish this chapter by considering two tidal sites. The first is La Rance, France where the world's first major tidal electric power plant was completed in 1966. The second is Passamaquoddy (U.S.–Canada) which has been considered as the site for a power plant for almost half a century.

TABLE 8-1.
Characteristics of Likely or Producing Tidal Power Sites

Site	Average Tidal Range	Basin Area	P_{max} (10^3KW)	E_{AN} (10^6KWH)
Minas–Cobequid North America	35.1 ft.	300 sq. mi.	19,600	172,000
White Sea U.S.S.R.	18.5 ft.	772 sq. mi.	14,000	122,000
San Jose Argentina	19.4 ft.	289 sq. mi.	5,760	50,500
Shepody North America	32.2 ft.	452 sq. mi.	2,490	21,800
Passamaquoddy U.S.–Canada	18.1 ft.	101 sq. mi.	1,750	15,300
Severn England	32.2 ft.	27 sq. mi.	1,490	13,100
Mezen Estuary U.S.S.R.	21.6 ft.	54 sq. mi.	1,330	11,600
La Rance France	27.6 ft.	8.5 sq. mi.	343	3,000

8.4 LA RANCE

The world's first major tidal electric power plant was completed in 1966 at La Rance in France, on the English Channel. Figure 8-3 shows water flowing out from the La Rance tidal power plant to the channel and the ocean. The capacity of the La Rance plant is 240,000 KW. Power is generated on incoming as well as outgoing tides. The maximum flow rate is near 650,000 cfs. Tidal ranges vary from 20 to 47 feet. La Rance uses a single-pool concept, and has 24 generator units. The dam across the

FIGURE 8–3. The Rance River tidal power plant. The only major tidal power plant in the world (in 1973) is at La Rance on the west coast of France. This 240,000 KW plant harnesses what appears to be "free energy." But the reader must realize that the building costs—$100 million in this case—make the energy anything but free. The cost of harnessing "free energy" must always be considered. In this photograph, the water which had entered with the rising tide, and had been trapped in the river basin to the left, is now being released through the dam, in which it turns turbo-generators. (*Courtesy Electricite de France.*)

narrow estuary mouth is a 2,300 foot long hollow reinforced-concrete structure. The cost of the plant is about $100,000,000.

The 240,000 KW La Rance plant can be called moderate in size. It makes a contribution to France's generation capacity (about two percent) and yet it is far smaller than some proposed plants. It must be considered an experimental plant if for no other reason than that it is the first major tidal plant, and many agencies considering tidal plants will no doubt be watching it closely.

Availability of a maximum flow rate and a tidal range suggests that we try to apply Equation (3-4) for the generation of hydroelectric power. Using $Q = 650,000$ cfs, $H = 30$ feet, and $\epsilon = 0.85$, we obtain $P_t = 1,400$ MW. This is about six times the actual 240 MW capacity of La Rance. This difference reflects the fact that a tidal plant cannot produce maximum power for very long, so water is stored to produce less power for a longer period.

Why does the La Rance plant use 24 turbines instead of one large turbine? There are a number of good engineering reasons for this choice. First, the effective head is greater for the smaller machines. A very large machine, comparable in diameter to the tidal range, would lose much of its effectiveness. Second, if one unit fails, the capacity of the plant decreases by only four percent whereas if you have only one turbine, failure means one hundred percent loss in capacity. It may be argued that the probability of having a failure is greater with 24 machines. This is true but if the probability of a single failure is small, the probability of a number of failures is *very* small. Third, small turbines are easily removed for inspection, maintenance, etc. Fourth, this large number of machines permits some use of assembly–line production.

The length of the dam (2,300′) is quite small. This factor is very important in holding down capital costs. A wide estuary with a narrow mouth is an important feature of an economical site.

Finally, we consider the plant cost ($100,000,000). Dividing by the capacity gives a KW cost of $400. This is competitive with most hydroelectric plants.

8.5   PASSAMAQUODDY

The Passamaquoddy site is on the U.S.–Canadian border (Maine–New Brunswick). Almost 50 years ago an American engineer, Dexter P. Cooper, proposed a two–basin scheme. The financial catastrophe of 1929 was too much for this proposal and it died, only to be resurrected in somewhat different form by a federal public works program in 1935. An end of Congressional appropriations terminated the study two years later, though not before valuable data for later studies had been gathered.

In 1948 an International Joint Commission was set up to review the concept. Its report in 1950 led to additional work from 1956 to 1959. Following these studies, the Department of the Interior developed plans for a 1,000 MW plant. The plan envisages 100 10 MW turbogenerators and seven miles of rock–filled dams (some in depths of 125 to 300 ft.). Many of the engineering problems have no precedent. After five years of study, ending in 1961, the IJC reported that the project was economically unfeasible by a wide margin, that the project could not produce power at a price competitive with the price of power available from alternative sources.

But Passamaquoddy does not die easily. President Kennedy asked the Secretary of the Interior to reopen the study. The subsequent study by the Interior Department found that the project could be feasible if operated as a peaking power source.(3) The proposed configuration for the plant is shown in Figure 8-4. Passamaquoddy Bay forms a 101-square mile high pool with the smaller Cobscoo Bay acting as the low pool. The total project would require seven miles of dams, 160 water gates for emptying and filling pools, and a number of maritime locks.

Capital cost of the Passamaquoddy Project was estimated, in 1963, at about $750,000,000, with energy costs about four mills/ KWH. Inflation would certainly have greatly increased those costs if the project were initiated today.

After the Interior Department report was released in 1963, the investor-owned utilities in the area retained Charles T. Main, Inc., a Boston engineering consulting firm, to study the project. The Passamaquoddy project would of course be a government project, and would be a competitor of the private utilities. Main's analysis indicated that tidal power would actually be more costly than that from alternative sources. There is no clear answer as to which party is right. Each seems right from its own viewpoint; the feasibility argument is at a standoff.

Why is there today a tidal power plant at La Rance, but not at Passamaquoddy? The public power-private power issue is no doubt a major factor, but one must also be impressed by the differences in the engineering challenge at the two sites. We have seen in the previous section that, in many ways, the technical problems at La Rance were *relatively* simple.

At Passamaquoddy the technical problems are far greater than at La Rance. In addition, the presence of the Passamaquoddy site on an international boundary introduces a whole new range of problems.

In asking for a new study at Passamaquoddy, President Kennedy observed, "each day a million kilowatts of power surge in

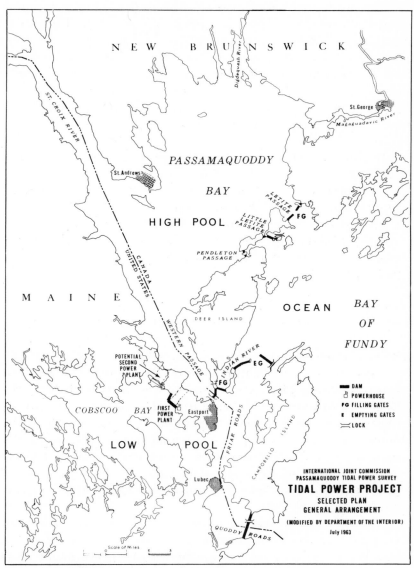

FIGURE 8—4. Map of the Passamaquoddy Project area.

and out of Passamaquoddy Bay." This is true, but it is not a complete justification for development. As we saw in Chapter 4, every project must be economically justified and financially feasible. Whether Passamaquoddy or any of the world's other great tidal sites someday meet these criteria is a question for the future.

REFERENCES FOR CHAPTER 8

1. G.D. Friedlander, "The 'Quoddy Question—Time and Tide," *IEEE Spectrum*, September, 1964, pp. 96–118.
2. National Academy of Sciences—National Research Council, *Resources and Man*, W.H. Freeman and Co., San Francisco, 1969.
3. U.S. Department of the Interior, "The International Passamaquoddy Tidal Power Project and Upper Saint John River Hydroelectric Power Development," Report to President John F. Kennedy, July, 1963.

GENERAL READING FOR CHAPTER 8

1. W. Ley, *Engineers' Dreams*, The Viking Press, New York, 1958.

    This is a very fascinating account of a number of fantastic projects that men have visualized in different parts of the world. Some are forerunners of great future projects. Some are already obsolete or too incredible to consider seriously—perhaps. Included are projects on solar, geothermal, tidal, and wind power; harnessing of the Mediterranean; construction of huge lakes in Africa; taming volcanoes, and many more.

2. A. Defant, *Ebb and Flow*, The University of Michigan Press, Ann Arbor, Michigan, 1958.

    This is a short book on the tides: what causes them, how they are measured, their force, and more. This book does not contain material on tidal power projects. It is not highly technical, and it can be read with relative ease.

3. A.J. Ippen, *Estuary and Coastline Hydrodynamics*, McGraw-Hill Book Co., New York, 1966.

This is a highly technical, detailed study of ocean waves, tides, and estuaries. It is appropriate only for the serious student of ocean–water motion and its effects. It has no information on tidal electric power.

4. T.J. Gray and O.K. Gashus, Editors, *Tidal Power*, Plenum Press, New York, 1972.

This is a collection of technical papers on the use of the energy of the tides. It is not appropriate for the general reader, but it may be of substantial interest to the technical student.

PROBLEMS FOR CHAPTER 8

General Problems

8.1. Suppose that a tidal basin has a surface area of 10 square miles and a tidal range of 25 feet. Assume that the basin is filled and the tide then recedes to its low value. What energy is potentially available in the impounded tidewater? Express your answer in ft.-lbs. and the equivalent KWH. (5.42×10^{12} ft.-lbs.; 3,670,000 KWH)

8.2. In Equation 8-4, P_{max} is expressed in ft.-lbs./sec. Find expressions for P_{max} and E_{AN} in KW and KWH respectively.

8.3. Find P_{max} and E_{AN} for the example in Problem 1; express in KW and KWH.

8.4. Draw a tidal–power time cycle such as that shown in Figure 8-2, except that the power generated is to be constant. Sketch qualitatively the upper and lower pool levels.

Advanced Mathematical Problems

8.5. For the student acquainted with simple electronic circuits, a fascinating analogy exists between a halfwave rectifier circuit with a parallel resistor–capacitor load and the two–basin tidal scheme. (See, for instance, R.J. Smith, *Circuits, Devices, and Systems*, John Wiley and Sons, Inc., New York, 1971, p. 396.) Develop the analogy by showing what elements of the two systems are analogs. How does the analogy fail?

Advanced Study Problems

8.6. Make a comparative study of tidal and hydroelectric power plants. How are they similar, and how different? Discuss as many properties or factors as you can.

8.7. Read "Islands Afloat," from *Engineers' Dreams* (Reference 1). Discuss some of the reasons that this dream has never been brought to fruition.

8.8. Determine through a search of recent periodical literature the status of tidal power plant development in the United States and elsewhere.

9

SOLAR POWER

*The sun, which passeth through pollutions
and itself remains as pure as before.*

*Francis Bacon
Advancement of Learning (ii)*

Energy from the sun gives life to the earth. Without it we would have no vegetation, no winds, no rain, no life. Earth would be a dead and barren planet. But of course the sun does warm the earth continuously, and man has always made use of the sun's energy. It is the ultimate source of all of our electric energy, with the exception of nuclear energy and, possibly, geothermal energy.

The solar power which falls upon the earth is an immense 177 trillion KW (1.77×10^{14} KW), which is 500,000 times the electric power capacity of the United States. Hence the available energy from the sun is far in excess of our foreseeable future needs. Can we find a way to harness this energy, though? That is the question we shall explore in this chapter. We consider three schemes which have been developed to some extent or which have been proposed.

1) Home systems
2) Orbiting space stations
3) Ground-based thermal plants

These examples do not exhaust the possible configurations, but they are representative of projects presently under consideration or development.

9.1 SOLAR-POWER DENSITIES

Table 9-1 lists the approximate solar-power density incident upon the earth, above the atmosphere as well as at the earth's surface. The energy from the sun near the earth but beyond the atmosphere is expressed in terms of a *solar constant.*(1) Even this presumably constant value varies slightly, depending on the condition of the sun (number of sunspots, distance from sun to earth, and so forth). The second row in Table 9-1 gives the solar-power density for a clear day with the sun directly overhead.(2) The third row accounts for the fact that the sun is not directly over-

TABLE 9-1.

Approximate Solar-Power Densities

Condition	Watts per sq. in.	Watts per sq. ft.	Watts per sq. yd.	MW per sq. mi.
In space near earth (solar constant)	0.90	130	1,170	3,630
On Earth, sun overhead; clear day	0.64	92	830	2,570
On Earth, sun overhead; clear day, 8 hour average	0.54	78	710	2,180
On Earth, 40° Latitude; clear day, 8 hour average	0.37	53	480	1,490

head very long; the row entries, therefore, average the power over eight hours of the sun's apparent motion. Finally, the fourth row takes account of the fact that, over a year's time, the sun varies considerably in its noontime angle from an overhead position. The figures given assume a latitude of about 40°.

Still other factors, which have not been considered, will decrease the average available power still further. These include cloudy days, air pollution or smog conditions, averaging over night-time hours, and the conversion efficiency of the proposed solar power plant system. All such factors must be examined in the initial study of a potential system at a given site.

The question of conversion efficiency is particularly important. It may range from a value well below 10 percent to possible values above 75 percent in some future systems. Even with relatively low conversion efficiencies, the densities given in Table 9-1 are large enough to encourage us to consider both small single-dwelling solar plants with a few square yards of collection area and large central power stations with many square miles of collection area.

9.2 HOME SYSTEMS

According to Table 9-1, on a clear day one-half KW or more may be incident on each square yard in a middle-latitude region. A typical home may have available perhaps 10 to 100 square yards (on a roof, for example) which would permit collection of about 5 KW to 50 KW of power averaged over eight hours. Experimental and operating home solar energy plants, in a variety of configurations, have been built for many years in many parts of the country.(3) Usually, the plant produces warm or hot air or water. (Figure 9-1.) It is conceivable that small solar electric plants for individual homes may become practical in the future.

Most people have observed that a garden hose which has been left out in the sun emits warm or hot water for a few seconds when the faucet is turned on, until all the water standing in the hose has passed through. Perhaps the simplest of "solar plants" employs a suitable length of black hose (to absorb radiation) placed in the

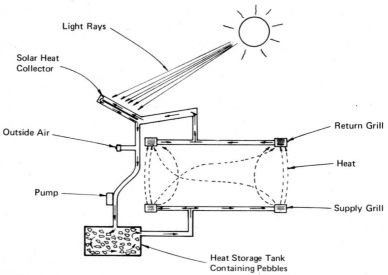

FIGURE 9–1. A solar heating system for a house. This illustration shows how energy from the sun is used to heat air in a circulating system. The air is pumped to a heat–storage tank filled with small stones.

open sun. Such a system will produce a bit of warm water, depending on the weather conditions, which could be used for bathing or washing clothes or dishes. System cost is the cost of an old black hose.

The next step is to build a box to hold the hose, and to cover it with one to three sheets of glass which allow solar radiation to enter but trap most of the infrared heat energy produced by the sun. This is essentially the *greenhouse effect*(4) which is used to grow plants, and which also operates on a global scale with carbon monoxide acting as an atmospheric trap for infrared energy. This system will cost quite a bit more money and produce quite a bit more heated water for the same collection area. This is an excellent example of a *capital–intensive* system, in which the initial investment is very important to the success and effectiveness of the system, and operating and maintenance costs are of little importance. In this case, of course, there is no "fuel" cost, but it will be necessary to perform occasional maintenance tasks such as cleaning the glass, removing corrosion from the pipes, and painting the system. Such costs should be a small fraction of the energy costs for this system.

The system above can be modified almost indefinitely, increasing the efficiency and the cost with changes in the shape of the collector, its orientation with respect to the sun, the convective losses of the water pipes (formerly black hose), and many more. Practical working solar plants are sold in many countries. Japan has a number of models. No doubt this form of energy harnessing will become increasingly attractive, particularly in remote areas, and in areas where the cost of fuel is high.

An obvious problem with such a scheme, as with any land-based solar scheme, is that it is highly sensitive to the availability of sunlight. Darkness and clouds render the system something between partially and totally ineffective, depending on whether or not heat storage is built into the system. Heat storage may be possible but it will mean additional expense. As a rule, of course, the more useful or versatile the system is, the greater its cost.

9.3 SPACE SYSTEMS

In the previous section we saw that small local solar plants have not been widely adapted to electric power generation, and are also sensitive to the presence of sunlight. A highly imaginative scheme has been proposed by Glaser(5) to convert solar power to electric power on a continuous basis.

Glaser proposes the use of a huge satellite, far above the earth's surface, in a *synchronous orbit*. That is, the satellite would be at such a height (22,300 miles) that its speed of rotation around the earth would be just equal to the speed the earth rotates. In this way the satellite is always over the same place on earth. A very large solar panel would be used to collect the sun's energy and convert it into electricity. One way to accomplish this task is to use a solid-state semiconductor diode, a relative of the transistor, which converts solar energy to electric energy. (See also Reference 6.) Such solar cells have an efficiency of about 10 to 15 percent. The electric energy generated by these solar cells would be used to drive a powerful microwave beam which would be transmitted to earth. The scheme is outlined in Figure 9-2.

This scheme has four critical components: 1) the solar panel, which must collect the sun's energy efficiently and convert it into electricity; 2) the satellite controller, which must maintain the po-

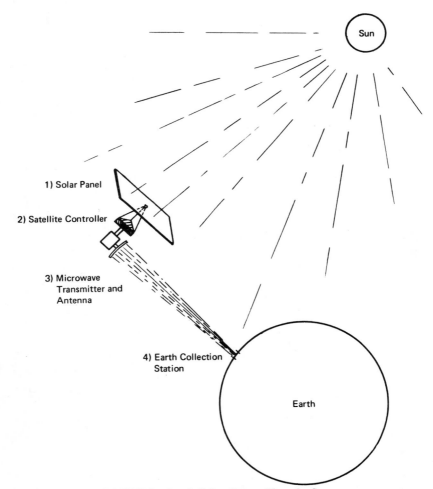

FIGURE 9-2. A Solar Power Plant in Space.

sition of the satellite panel with respect to the sun and antenna with respect to the earth; 3) the microwave transmitter and antenna, which must change the available electric energy into microwave electromagnetic energy and transmit it in a highly concentrated and directed beam back to earth, and 4) the earth's receiving antenna site. Each of these critical elements presents serious technological problems which must be solved before the scheme can be implemented. We consider each of these problem areas briefly.

First we consider the solar collection panel. Such devices have been used for years to provide electric power for satellites. Power to run a satellite may allowably be very expensive, but commercial power must be competitive in the marketplace. So we must expect any proposed system to be able to produce energy at perhaps six to 10 mills per KWH. Economy of scale effects suggest that we will probably wish to build very large solar plants, of perhaps 10,000 to 25,000 MW. The cost of solar cells is presently about $400 per square foot.

Example 9-1

Consider a plant of the type described above, with a total overall efficiency of 8 percent, a desired power output on earth of 12,000 MW, and a plant factor of 80 percent. Find the cost of energy due only to the capital costs of the solar cells.

An output power of 12,000 MW and an efficiency of 8 percent requires an input power of:

$$\frac{12,000}{0.08} = 150,000 \text{ MW (solar)}$$

Referring to Table 9-1 we see that the solar power intensity near the earth is about 3,000 MW/sq. mi. The required collection area is 50 square miles, which means a square panel about seven miles on a side, or a round panel of diameter about four miles. Solar cells now cost about $400/sq. ft. or about $10,000,000,000 per square mile. Hence the total cell capital cost is approximately $500,000,000,000 (500 billion dollars). Assume we amortize the capital (if we could find it) over 50 years at 6 percent. Referring to Table 4-1 we see that an annual payment of 6.3 percent, or

$31,500,000,000 is required. A plant factor of 80 percent means the number of hours per year the plant is used is:

8,760 × 0.8 = 7,008 hours

Hence the total energy produced is:

7,008 × 12,000,000 KW ≅ 84,000,000,000 KWH

Finally, to find the energy cost per KWH due to the capital cost of the panels alone we divide capital cost by energy

$$\frac{\$31,500,000,000}{84,000,000,000} = \frac{\$0.375}{KWH} = 375 \text{ mills/KWH}$$

It is clear from the above example that solar panels are prohibitively costly and inefficient at this time. It will be necessary to reduce costs drastically or increase efficiency of panels before such plants can become financially feasible. It is not obvious at this time that this can be done, but there are some possible alternatives to present cells, such as organic cells, or vastly improved inorganic single-crystal semiconductors, which may in the future permit the cost-efficiency combinations necessary to justify commercial solar cell panels.

The reader will note that we have not, of course, completely analyzed the costs of orbiting 50 square miles of solar panels. The weight of such a panel might be 10,000 to 50,000 tons. Shuttle systems presently under consideration could build such space stations. The costs of orbiting and building are very difficult to estimate but they would certainly be quite significant.

We consider next the problem of the position of the satellite. It would be necessary to provide constant control to maintain the correct synchronous orbit, as well as panel and antenna orientation. Such control would be essential and probably within the present state of technology.

A second problem related to satellite position is that for short periods during the year, near the equinoxes, the satellite is in the earth's shadow. That is, it does not see the sun. During these periods it would not be able to transmit power back to earth. It has been suggested that a solution to this problem is to place two satellites in orbit such that one is always in the sun. The theory is that power would then never be interrupted at the receiving site. This approach appears to have a serious economic flaw. Most of the capital cost of the system would probably be in the satellite system, rather than the ground receiving station. Hence, once a satellite is orbited, it is highly desirable to use it as much as possible. We would no doubt want to have both satellites transmitting energy back to earth, using this as baseload. The problem then is that when one satellite is shadowed, the baseload drops by 50 percent. Hence the problem has not changed significantly from that of the single–satellite problem.

The third problem concerns the microwave transmitter and antenna. No single microwave transmitter tube can handle anywhere near 10,000 MW, but 10,000 of them might generate 1 MW each with only a fairly minor extension of the state of the art. The antenna would need to be about one mile in diameter to concentrate a beam which would radiate onto an area on earth of perhaps 20 square miles.

Example 9–2

If 12,000 MW are transmitted to an area of 20 square miles on earth, what is the power density of the beam in watts/square inch?

The area in square inches of 20 square miles is:

$$20 \times 5{,}280^2 \times 12^2 \times 80{,}389{,}792{,}000 \text{ sq. in.}$$

The beam intensity is approximately:

$$\frac{12{,}000{,}000{,}000}{80{,}000{,}000{,}000} = 0.15 \text{ watts/sq. in.}$$

Microwave power intensities near that of the above example may be somewhat but probably not severely damaging to living things. It would no doubt be desirable or necessary to restrict human access to the receiving antenna region, to restrict aircraft flights through the beam, and to have a highly reliable way of turning off the beam if it were misdirected to some other region on earth.

Finally, we turn to the problem of the earth's receiving station. The station would act like a huge antenna, made up of dipoles connected to very high efficiency solid state rectifiers. The rectifiers would change the incoming microwave (alternating current) signal to a direct current (DC) signal which would be fed to transmission lines and eventually transmitted to appropriate load centers. Research on networks such as this has been initiated, but it is not clear at this time exactly how such a station would be developed.

A solar station in space is certainly a most exciting proposal. It also is one with many difficult technological challenges. Whether the technology and the economics will ever justify such development is impossible to predict.

Meanwhile, as this approach is reviewed and studied, our attention turns back to earth to another proposal for harnessing solar power.

9.4 A LARGE GROUND-BASED SOLAR PLANT

If we wish to avoid the problem of building and orbiting a huge satellite, we can of course think of using a ground-based station. That avoids the problems discussed in the last section. In their place it introduces a new series of problems, primarily concerned with loss of sun at night, weather, and atmospheric attenuation.

One possibility is to put the solid-state solar panel of the last section in a ground configuration. However, this system is far too costly at this time, as we saw in Example 9-1, and is also rather inefficient at this time. The search turns to some alternative scheme which might offer hope for acceptable efficiency and cost while avoiding the problems of a space station. Such a scheme has been proposed by Aden Meinel of the University of Arizona.(7)

The Meinel scheme (Figure 9-3) calls for a large network of copper pipes containing liquid metal (Sodium or a mixture of Sodium and Potassium) to be heated by the sun. The pipes would carry heat at about 560° C to a large combination heat storer and exchanger. This device plays the dual role of storing heat during times when the sun is not on the pipes and transferring heat to a steam generator for interface with a conventional steam power plant.

The group of heat-carrying pipes and associated reflector panels

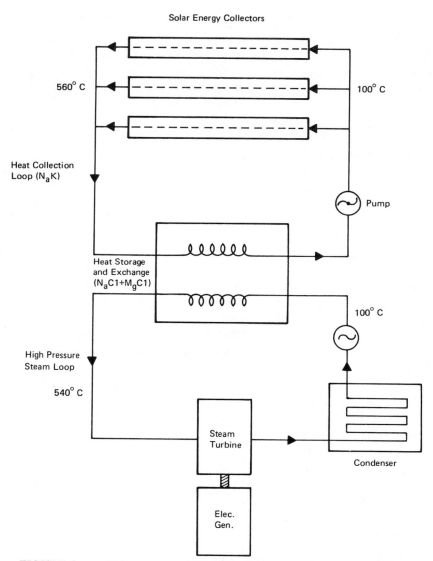

FIGURE 9–3. Meinel Scheme for Land-Based Solar Power Generation.

at the top of Figure 9–3 would actually be a "solar farm" in the desert, many square miles in area. The hot liquid metal would be pumped to very large cylindrical underground tanks which contain an appropriate mixture of salts (such as NaCl and MgCl). The hot metal would transfer its heat to the salt in the tank through a heat exchanger. Heat would then be stored within the tank, until it is needed, for as long as weeks.

In the lower loop of Figure 9–2, water is pumped into the heat storage chamber; it leaves as high pressure steam. The steam drives a conventional steam–turbine electric generator to produce electric energy. The steam leaving the turbine must of course be condensed for the same reasons as given in Chapter 4. The Meinel scheme uses a thermal cycle, whereas Glaser's space station employs a direct conversion scheme from electromagnetic radiation to electric energy.

The important implication of Meinel's thermal cycle is that condenser cooling water is required. This would require major sources of cooling water in the arid deserts of the Southwest. The waste heat could be used to desalinate sea water from the Pacific Ocean or from the Gulf of California, however. This would produce a valuable byproduct. The combination of electric power and fresh water in desert areas would open up some rather extraordinary opportunities for new development. The implications or effects of such development are not clear at this time. We will discuss below some of the possible effects. First, let us turn to the question of dimensions of our plant.

It is clear from our study of solar plants in space that solar energy is quite diffuse, and a given power plant size will require a relatively large surface area. We saw in Example 9–1 that a plant of 12,000 MW size, operating at eight percent efficiency, would require close to 50 square miles. On earth the area would be even larger because of the loss of energy through the atmosphere, during night time, and on cloudy days.

Meinel has made a rough calculation of the amount of land required to produce 1,000,000 MW, which is about half of the projected U.S. power demand in the year 2000. The result is in excess of 13,700 square miles, or about 14 percent of the U.S. desert area.

A number of problems arise. First, do we want to use this amount of land to "farm" energy from the sun? One answer is that we presently cultivate far more land for farms which grow food, at a much lower efficiency level. A second answer is that while this approach may not be ideal, it may be better than any of the alternatives, and the sacrifice of land should be made. A third answer, offered by many conservationists, is that the sacrifice is too great.

A second question is whether the system would have significant effects on the weather conditions of the Southwest. The answer is not really known.

To get a better idea of the magnitude of the project, consider Figure 9-4.(7) In this figure we see the proposed Meinel solar

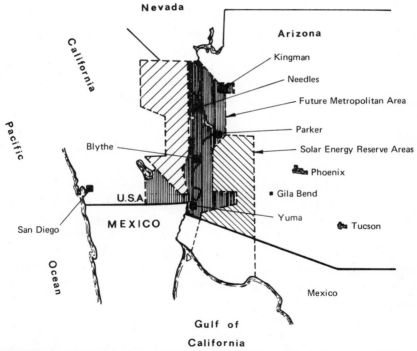

FIGURE 9—4. Map showing possible areas for the proposed National Solar Energy Reserve and the resulting Colorado River Metropolitan Area.

farm area, which is located in the desert country of the Southwest. Large areas on either side of the Colorado River would be "farmed" for solar power. Along the river a greenbelt area would no doubt develop, with major new industry, housing, irrigation, and vegetation. Industry would be attracted to the new source of power, and it might even be necessary to keep energy costs down. People would be required to operate the power system and new industries. The Meinels have estimated that as many as one to two million people would be needed to operate and provide services related to a one–million megawatt system.(8)

A development of this magnitude certainly would have an effect on local and possibly also on regional weather conditions. Such effects would be observed as the project grew in magnitude. Obviously any change which altered the presence of clear skies would be undesirable or unacceptable. It might be necessary to restrict the use of gasoline–driven vehicles in the vicinity because of the resulting smog and its effect on the solar collectors.

Another factor which would have to be considered is the interaction with other major new projects in the area, such as the proposed Salton Sea geothermal project. (Reference 5, Chapter 7.) Both of these projects involve extensive desalination programs. Another possible area of interaction is the so–called Four Corners fossil-fuel development, which at this writing is generating unexpectedly large amounts of air pollutants.(9) The Mojave plant, fed with coal from Four Corners, is located near the northern end of the proposed future metropolitan area shown in Figure 9-4.

If the Meinel scheme is developed, the interrelated questions of land use, weather changes, and related projects would no doubt have to be reconsidered at each stage of the basic development.

Another consideration is that of raw materials. Many of the needed materials are abundant; others are much less so. It remains to be seen exactly what kinds of materials, and in what quantities, will be required.

Finally, there is a question concerning what would be done if the region experienced a long string of cloudy days. There would be no backup system, apparently, to bail out the residents who depended upon solar energy. The result of cloudy periods might be significant cutbacks in power. You may wish to recall that the proposed scheme, if fully developed, would represent more than 50 percent of the nation's power capacity. Reliance on such a system would require us to consider temporary power shortages and curtailment; we would probably also need new systems of storage of energy.

These are just a few of the problems which must be faced and solved. The approach in this case almost certainly will be cautious and relatively slow. Some additional initial studies are necessary. They will probably be followed by a small pilot plant, and later a significant commercial plant if no major obstacles arise. Only very slowly will we need to decide whether we have the ability, the economic feasibility, and the desire to cover much of our desert land with solar collecting plates.

9.5 CONCLUSIONS

The major lessons of this chapter are that the energy of the sun is immense, and that it is very diffuse. The former offers us an extraordinary opportunity to obtain electric energy without the sacrifice of depletable fuels. The latter leaves us with the sobering challenge of trying to find an acceptable way to harness this diffuse source. In the years between 1970 and 1990 we should find out if we are able to harness the sun. If we are, we shall then have to decide if we wish to do so.

REFERENCES FOR CHAPTER 9

1. J.I. Yellott, "Solar Energy," in *Standard Handbook for Mechanical Engineers* (Baumeister and Marks, Editors), Seventh Edition, McGraw–Hill Book Co., New York, 1967.
2. N.C. Ford and J.W. Kane, "Solar Power," *Science and Public Affairs*, October, 1971, pp. 27–31.
3. D.S. Halacy, Jr., *The Coming Age of Solar Energy*, Harper and Row, New York, 1973.

4. J.S. Sawyer, "Man–made Carbon Dioxide and the 'Greenhouse Effect,'" *Nature*, Vol. 239, September 1, 1972, pp. 23-26.
5. P.E. Glaser, "Power from the Sun," *Mechanical Engineering*, March, 1969, pp. 20-24.
6. R.J. Smith, *Circuits, Devices, and Systems*, John Wiley and Sons, Inc., New York, 1971, pp. 463-464.
7. A.B. Meinel, "A Proposal for a Joint Industry–University–Utility Task Group on Thermal Conversion of Solar Energy for Electrical Power Production," University of Arizona, April 27, 1971.
8. A.B. Meinel and M.P. Meinel, "Is It Time for a New Look at Solar Energy?" *Science and Public Affairs*, October, 1971.
9. A. Wolff, "Showdown at Four Corners," *Saturday Review of the Society*, June 3, 1972, pp. 29-41.

GENERAL READING FOR CHAPTER 9

1. "Power from the Sun" from General Reading Citation 1 in Chapter 8.

    This chapter of W. Ley's book on the dreams of engineers is a very interesting review of some of man's efforts to harness the sun.
2. B. Chalmers, *Energy*, Academic Press, New York, 1963.

    This partly technical book has a considerable amount of material on solar energy as well as most of the other forms of energy discussed in the present work. Most of the material can be understood by the average reader.
3. F. Daniels, *Direct Use of the Sun's Energy*, Yale University Press, New Haven, 1964.

    This is a complete and competent discussion of all the major aspects of solar energy. It includes some history, theory, and a good many applications. It is written at a relatively low level of difficulty, and it contains many references.
4. J. Hoke, *Solar Energy*, Franklin Watts, Inc., New York, 1968.

    This simple introductory work on solar energy emphasizes pictures and elementary explanations. It illustrates a number of examples of use of the sun's energy.

5. D.D. Halacy, Jr., *The Coming Age of Solar Energy*, Harper and Row, New York, 1973.

This is a good contemporary discussion of solar energy. It contains detailed material on a number of existing or proposed processes, including the Glaser scheme for a space station and the Meinel scheme for farming the desert. The level of difficulty is not high.

PROBLEMS FOR CHAPTER 9

General Problems

9.1. A home solar power plant is set up at a location in latitude 40°. Assume that, because of 24–hour averaging and some cloudy days, the average solar power density is 100 watts per square yard. A 25 square yard panel is set up on the roof. If the conversion efficiency of the system is 10%, what is the available average power, energy per month, and the value of that energy at a rate of $.03 per KWH? (250 watts, 180 KWH, $5.40)

9.2. Consider the previous example. If the total system efficiency were increased to 100%, would the plant become feasible?

9.3. Suppose that the Meinel scheme were fully implemented, with the development of 1,000,000 MW of power. About how much capital could we afford to spend on such a facility?

9.4. Indicate some additional problems which might arise in the implementation of the Meinel scheme.

9.5. Which states have an area roughly equal to that of the proposed Meinel solar farm?

Advanced Mathematical Problems

9.6. What is the energy cost due to the solar panel if the efficiency is increased to 70% and the cost reduced to $2.00 per square foot? (The plant factor is 90% and the cost is amortized over 40 years at 5% interest.)

9.7.   Write an equation for the solar power density as a function
       of the angle of incidence of the sun's rays on the earth. As-
       sume that when the sun is directly overhead, the density is
       Po. What do you have to assume to write this equation? How
       could you alter the equation to compensate for this assump-
       tion?

Advanced Study Problems

9.8.   Read Reference 8 in this chapter and Reference 5 in Chap-
       ter 7. Which factors are in conflict, if any? Which goals are
       common to both? Are there ways in which the projects
       could either co–exist or perhaps enhance each other?
9.9.   Build a small solar plant of any form. Test it, and, if pos-
       sible find a way to measure its efficiency.
9.10. Read Reference 3, particularly Chapter 1, for one author's
       sense of the sun's role in providing man with energy. Discuss
       Claude Summers' "thermal ceiling." How is this concept re-
       lated to concentrating people in relatively small areas?

10

NUCLEAR FUSION

The wrong of unshapely things is a wrong too great to be told,
I hunger to build them anew and sit on a green knoll apart,
With the earth and the sky and the water, remade,
  like a casket of gold,
For my dreams of your image that blossoms a rose
  in the deeps of my heart.

> William Butler Yeats
> *The Lover Tells of the Rose in His Heart*

There is perhaps no potential source of electric energy yet conceived that offers greater promise or greater challenge than nuclear fusion. Of fusion we can say:

1) Its fuel sources are essentially limitless.
2) It is inherently less dangerous than fission power.
3) Storage of fusion waste products is less difficult than storage of fission waste products.

4) We don't know how to make it work.

In this chapter we explore both the promise and the challenge of fusion. We have a second objective in this chapter. You will recall that a second objective of Chapter 9 on tidal power was to argue that energy is not "free" until it is harnessed, and thus it is not free at all. In this chapter our second objective is to show that it takes decades from the time a power generation scheme is proposed until it can be put to work in large scale.

10.1 NUCLEAR FUSION REACTIONS

There are two important nuclear reactions which involve the release of energy. They are *fission* and *fusion*. Fission, which was discussed in Chapter 6, is the splitting or division of a heavy atom (or isotope) into two atomic (or isotopic) fragments of nearly equal mass, plus some radiation products. Fusion is the joining of two light nuclei with a release of energy. In this chapter we discuss how this process may some day be harnessed to produce electric energy.

A number of fuels have been proposed for possible forms of fusion reactors. The most important are the two heavy isotopes of Hydrogen, Deuterium ($_1D^2$) and Tritium ($_1T^3$). Deuterium occurs quite commonly in heavy water, D_2O. It occurs in the ratio of about one part to each 6,500 parts of light (ordinary) water in the sea. Tritium is a radioactive (electron–emitting) isotope which does not occur naturally in any abundance. It is, however, the by-product of some possible reactions.

Deuterium is a highly attractive fuel not only because it is abundant in sea water, but also because it is relatively easy to separate from light sea water. Heavy water has a mass more than five percent greater than that of ordinary water. This is far greater, for example, than the mass difference between U–235 and U–238, which must be partially separated in a crucial stage of nuclear fission fuel processing.

Many possible light nuclei reactions lead to fusion energy. One of the most important is presented here as an example.

$$_1D^2 + {}_1D^2 \rightarrow {}_2He^3 + n + 3.2\text{ MeV}$$

$$_1D^2 + {}_1D^2 \rightarrow {}_1T^3 + p + 4.0\text{ MeV}$$

$$_1D^2 + {}_1T^3 \rightarrow {}_2He^4 + n + 17.6\text{ MeV}$$

$$_1D^2 + {}_2He^3 \rightarrow {}_2He^4 + {}_1H^1 + 18.3\text{ MeV}$$

He represents Helium, n a neutron and p a proton. This chain of reactions can be summarized as follows:

$$6_1D^2 \rightarrow 2\ {}_2He^4 + 2p + 2n + 43.1\text{ MeV}$$

The byproducts are $_2He^4$, a stable isotope of Helium; two protons, which are simply nuclei of the principal Hydrogen isotope, $_1H^1$; and two neutrons. The 43.1 MeV of energy may seem small compared to the 200 MeV released in the fissioning of a heavy isotope like $_{92}U^{235}$. However, the energy released per mass of fuel is actually greater for the fusion case.

One important problem with the reaction above is that the D–D reaction rate is fairly low. An alternative cycle takes advantage of the greater reaction rate of D–T.

$$_1D^2 + {}_1T^3 \rightarrow {}_2He^4 + n + 17.6\text{ MeV}$$

$$n + {}_3Li^6 \rightarrow {}_2He^4 + {}_1T^3 + 4.8\text{ MeV}$$

where Li is the element Lithium.

The D–T reaction is more challenging to harness technologically in some ways and less challenging in other ways. The net result is that it is likely that a D–T reaction will be developed first, but that the D–D reaction will in time supplant it, assuming, of course, that fusion power is someday harnessed.

10.2 CRITERIA FOR NUCLEAR REACTIONS

The fuel in a fusion reactor is in the form of a gas. The reactions indicated in the previous section require extremely high temperatures, of perhaps 50,000,000° K to 200,000,000° K. (The sun is a lukewarm 20,000,000° K by comparison.) The reason for these high temperatures is that the nuclei are identically charged, and therefore have a tendency to repel each other. Only if the kinetic energies of the nuclei are extremely high will they be able to approach close enough to fuse. At the very high temperatures indicated above, atoms are stripped of their orbiting electrons. The gas becomes ionized—separated into positive and negative charged particles. In this state the gas is called a *plasma.*

In addition to the requirement of high temperature, a reactor which is to produce net power must have an acceptable combination of particle density, or nuclei per cubic centimeter, n, and particle confinement time t in seconds. The Lawson criterion requires that the product nt be equal to 6×10^{13} for the D–T reaction and 2×10^{15} for the D–D reaction. In Section 10.3 we discuss how plasma confinement may be accomplished.

The very significant factor of almost 30 times between the Lawson criteria for the D–T and the D–D reactions is the major reason why the D–T reaction is likely to be developed first.

Figure 10–1 is a plot of particle density versus temperature for a number of important plasmas.(1) Many of these plasmas should be familiar to the reader. The ionosphere, which is above the earth from an altitude of roughly 50 to 1,000 miles, is relatively cool and tenuous (non–dense). Even though its temperature is sometimes quite high relative to the earth's surface, its density is so low that an object such as a spaceship can pass through it without "burning up." A candle "flame," on the other hand, has a greater density and temperature, and it can burn some materials.

Temperatures and densities far greater than those for these last two examples are necessary for today's controlled fusion experiments and tomorrow's proposed fusion reactors. These regions are in the upper right hand corner of Figure 10–1. We turn now to the very difficult task of confining plasma particles with densities and temperatures in these regions.

10.3 PLASMA CONTAINMENT

It is not possible to contain a very hot plasma within a solid vessel made of steel, concrete, wood, etc. We would not have to worry about the containment vessel melting. The plasma density here is still so low that the vessel could absorb the heat without being destroyed. However, contact with the wall of the vessel would essentially quench the plasma, bringing the temperature of the gas down close to that of the vessel. In addition, local heating of the vessel would vaporize small fragments of it, producing con-

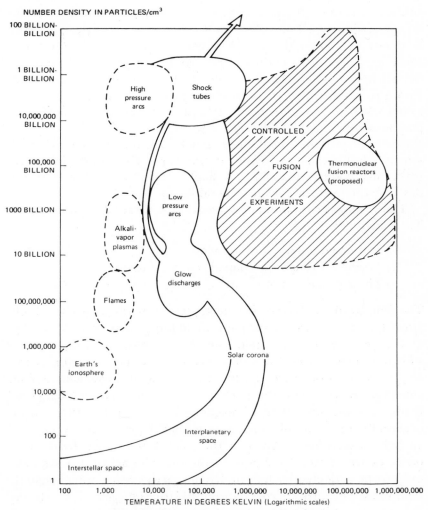

FIGURE 10–1. The World of Plasmas.

tamination which would inhibit the reaction process. Hence, it is critical that the plasma be contained away from the walls of the reactor.

The solution to containment lies in the fact that the plasma consists of electrically charged particles. Charged particles which move in a magnetic field experience forces which can be used to confine or contain the particles away from the walls of the reactor. We must design a kind of magnetic prison in which particles are held not by steel walls but by invisible magnetic field lines, like those produced by a child's toy magnet. Not just any prison will do. Escape must be very improbable. The basic task we have in fusion research today is to find a magnetic system which can contain a plasma of sufficient density and temperature for a sufficient length of time.

The physical theory of operation of the possible containment schemes is beyond the scope of this text. The interested reader may consult References 2, 3, 4, and 5. Here, we shall only briefly describe the most important general forms.

Three major containment schemes are presently under study. These include:

1) Stabilized mirror containment
2) Toroidal (doughnut-shaped) containment
3) Theta pinch containment

The stabilized mirror system being tested at the Lawrence Radiation Laboratory uses a magnetic field which is squeezed together at the ends so that the field lines come closer to each other, as shown in Figure 10-2. The effect is to cause the charged particles to reflect from or mirror back from the region in which the lines converge. In this way the particles continually bounce back and forth from one end of this "magnetic bottle" to the other. This is essentially the same phenomenon as that experienced by charged particles which are trapped in the earth's Van Allen Belt. These particles travel from one pole to the other in about 1.5 seconds, only to be reflected back again by the earth's converging magnetic field lines. This process continues until situations are such that any individual particle "leaks" out at one of the earth's poles.

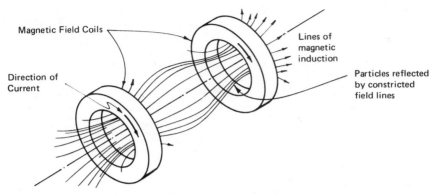

FIGURE 10—2. Magnetic Mirror Confinement Scheme.

In the magnetic bottle the plasma particles must be contained long enough to permit a significant amount of power to be generated. This may be a fraction of a second. One experiment at the Lawrence Radiation Laboratory yielded a density of n = 5 × 10^{13} particles/cm^3 and a temperature T = 80,000,000° K. However, the containment time was only about 0.3 msec, which is about one thousandth of the required time.

The toroidal configuration has been under study at the Kurchatov Institute of Technology in Russia and at Princeton University. The Russian device is called a Tokamak, the Princeton device a Stellerator. To date the Tokamak has had much greater success. In fact, the Princeton device has been converted to the Tokamak form. The principle is to force the charged particles to follow a circular path inside a toroidal structure. Figure 10-3 shows the basic Tokamak configuration, and Figure 10-4 is a picture of the Princeton Tokamak under test.

A theta pinch experiment is in progress at the Los Alamos Laboratory. The objective is to obtain a very rapid compression of an existing low–density, low–temperature plasma. The compression causes heating which leads to the necessary conditions for sustaining a reaction. Many other variations of containment schemes have been proposed. Some are under study at this time.(2-5)

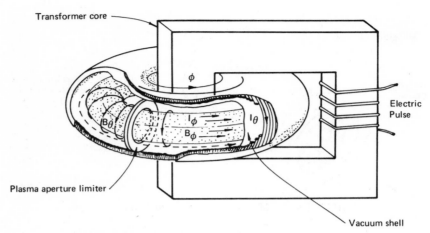

FIGURE 10—3. The Tokamak confinement scheme.

10.4 FUSION BY LASERS

Recently it has been proposed that fusion power may be obtained without magnetic containment through use of a high-powered *laser*. (6) The laser is a device which can focus a very high intensity light beam on a small region for a short period of time. In one scheme a Deuterium–Tritium pellet is dropped into a sphere or reaction chamber. When the pellet reaches the center of the chamber the very high intensity laser beam heats the pellet to fusion temperatures and a fusion reaction follows. Energy is deposited in the Lithium as heat. This is drawn off through a heat exchanger to produce steam for a conventional turbine.

There are two major problems facing this scheme. The first is that there are presently no lasers of sufficiently high power to trigger the reaction. The second problem lies in containing and harnessing the sudden release of energy which occurs in a shock wave form, and which might otherwise damage or destroy the containment vessel. The problems facing the laser approach seem at least comparable with those facing magnetic containment.

FIGURE 10—4. The Symmetric Tokomak, a controlled fusion research device in toroidal geometry, located at Princeton University Plasma Physics Laboratory and sponsored by the U.S. Atomic Energy Commission. The photograph suggests the toroidal, or doughnut–shaped, configuration of the Tokomak. According to *Survey of USAEC Program in Controlled Thermonuclear Research*, ". . . this machine was built to attack three main goals. The first, already accomplished, is to verify that such devices do produce hot, dense plasmas. The second goal is to determine the essential features of the confinement process and to evaluate the differences, if any, between Stellarator and Tokomak confinement. The third goal is to move beyond the present parameters, especially in the direction of higher ion temperatures." The *New York Times* of July 12, 1973 reports that the third goal is being successfully attacked by present research. The Princeton team has injected a stream of Deuterium atoms into the plasma in such a way that the pre–compression temperature of the plasma has been raised from about 3.5 million degrees to 4.5 million degrees. After compression, the plasma temperatures are about 10 million degrees without Deuterium injection, about 12.5 million degrees with Deuterium injection. Fusion reactors will require temperatures in the range of 180 million degrees. (*Courtesy Princeton University Plasma Physics Laboratory.*)

10.5 ELECTRIC POWER PLANT

So far our attention has centered on the fusion reaction itself. We turn now to the problem of using the thermonuclear energy, released in that reaction, to generate electric energy. Two basic approaches have been conceived, one using a thermal (steam) cycle, and one which converts reactor energy directly to electric energy. In addition, a number of proposals have been made for combining the fusion reactor with one or more secondary devices.

Figure 10-5 shows one form of a thermal cycle. A cross section of a toroidal reactor is shown on the right. In the center of the toroid the contained plasma moves in a vacuum. We have cut across the doughnut of the toroid; imagine that the plasma is flowing out of the surface of the page. Neutrons released in the reaction are absorbed by the liquid Lithium, which carries heat to a heat exchanger steam generator. The rest of the system to the right of the heat exchanger is essentially identical to the steam system studied in Chapter 5. Since neutrons react with Lithium to produce Tritium, which is needed in the D–T reaction, the Tritium is separated out from the liquid Lithium and reinjected into the reactor.

A direct energy conversion scheme, shown in Figure 10-6, has been proposed by Dr. Richard Post of the Lawrence Radiation Laboratory. A mirror containment device is used, as shown on the left. Some particles escape through the mirror on the right into a charge collection region. Electrons would be removed by electric collecting plates, much as they are in a vacuum tube. These electrons would then appear as an electric current in an external circuit.

A novel extension of the fusion reactor is the fusion torch, which would use the very hot plasma available from a reactor to burn refuse completely, reducing it to its basic elemental constituents. (1)

10.6 DEVELOPMENTAL PROBLEMS

At the present time we are in a stage of research directed at showing that acceptable containment is possible. Sometimes this is called a phase of scientific research. This is perhaps a misnomer.

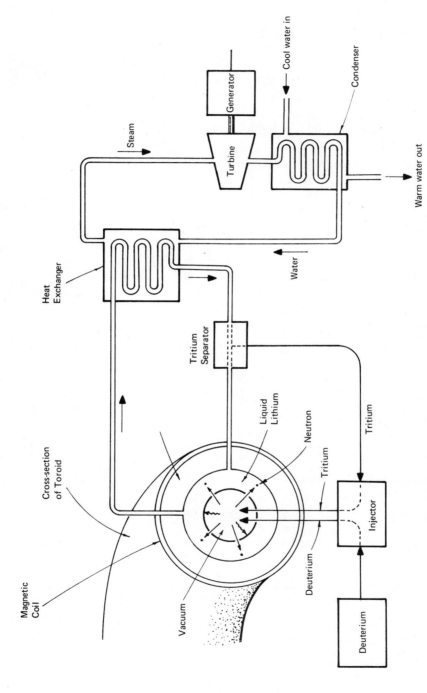

FIGURE 10–5. Fusion Reactor in a Thermal Generation Scheme.

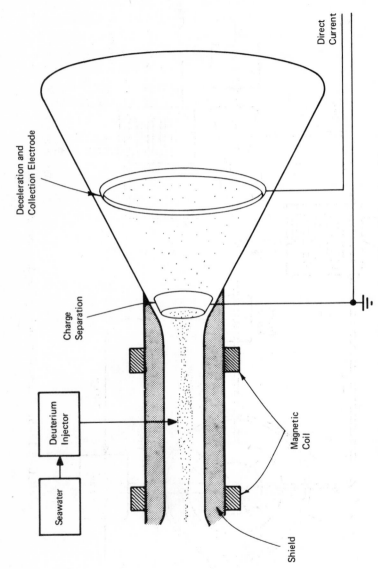

Deceleration and
Collection Electrode

Charge
Separation

Deuterium
Injector

Seawater

Magnetic
Coil

Shield

Direct
Current

FIGURE 10–6. Fusion Reactor in a Direct Conversion Scheme.

The scientific credibility of thermonuclear reactions has been established on the earth, and is demonstrated in the continued life of the sun and the stars. In a sense we are now trying to solve the first of a series of engineering challenges to achieve practical application of theory. Some experts (see, for example, Reference 4) believe that we may reach the minimum criteria for an energy-producing reaction in a few years. In recent years funding to this end by the Atomic Energy Commission has been close to $30,000,000. The percentage of total world fusion research funded by various nations in 1971 was:(7)

U.S.S.R.	37.5%
West Germany	16.6
U.S.A.	15.6
United Kingdom	7.0
Japan	6.1
Others	17.2

After we have reached the minimum criteria, we must then turn to a very wide range of serious application problems. We should anticipate that these will require many years to solve. The following partial list of major technical problems will suggest the magnitude of the task ahead. (8, 9)

1) Large, high-strength magnetic fields produced by super-conducting magnets will be needed. They must be developed and they must run at relatively low power levels.
2) Vacuum chambers that can withstand severe temperatures and neutron bombardment must be developed.
3) Tritium handling apparatus must be developed.
4) Technology must be developed for fueling and refueling reactors, recovering unburnt fuel, and handling waste products.
5) Schemes must be developed for start-up, shutdown, and load-following.

10.7 ENVIRONMENTAL QUESTIONS

Inevitably the fusion reactor is compared with its potential competitors. The breeder reactor, probably an advanced model, seems the most likely competitor. Solar power may compete. Geothermal power seems only an outside prospect at this time, considering the very large amounts of energy that might be produced by fusion.

The first fusion plant will probably be a thermal plant, with the same basic thermal pollution or heat addition problems as those of any other thermal plant. In time, however, a direct conversion scheme could lead to major reductions in waste heat release.

The major radioactive materials in fusion reactors are Tritium and the radioactive isotopes produced by neutron bombardment of the reactor's structure. The prospects of a major nuclear excursion seem very slight, since, if anything goes wrong in the reactor, the plasma will probably be quenched quite quickly by the reactor walls. If the reactor is catastrophically ruptured, the release of radioactive materials is far less serious by a factor of perhaps one million in fusion than in fission reactors.

The radioactive waste storage problem with a fusion reactor is much less severe than with fission reactors. Therefore, on environmental grounds, the fusion reactor seems quite attractive.

10.8 IS FUSION IN OUR FUTURE?

The history of fusion power is one of ebbing and flowing tides of optimism and pessimism. Today we ride the crest of a flow, if not a flood, of optimism. Many experts are beginning to predict in terms of "when" instead of "if." But most agree that the timetable for development is not short. One general schedule suggests five years to show that minimum criteria can be met. Another 10 years would be required to iron out the technical problems discussed in Section 10.6, and still another 10 years to develop the first full scale plant. That brings us up to the year 2000. Some say such a timetable is too optimistic; others say that is unduly pessimistic and that a major increase in research funds should be made, and would greatly shorten the timetable.

It has been suggested by some that we bypass the breeder reactor and proceed with all due speed to fusion. It is not likely that this will happen. Industry and government have essentially committed us to development of the breeder reactor. It is the surer device. We essentially know how to do it. Fusion is more of a gamble; the stakes are high, and the rewards are great.

It seems that we ought to wager more on the fusion scheme. The extra millions we spend today on feasibility determination and on the simultaneous attack on technical problems may yield huge gains in environmental protection as well as economic gain on some earlier tomorrow.

10.9 A FINAL THOUGHT

The problems we face in developing fusion power have one thing in common with any major new generation scheme. The timetable is long. It takes years to show technical feasibility, more years before the new scheme carries any large proportion of the load. Even though nuclear fission power was shown to be feasible over 30 years ago, and the first plants produced power over 15 years ago, nuclear power still produces less than four percent of our electric energy (1973). The coal and oil and nuclear plants we build today will be producing power 30 and 40 years from now, when fusion may be just coming on line. The lesson is that if we are to meet the apparent demand of the next two to three decades we will almost certainly have to do it with coal, oil, and gas along with Uranium and falling water.

REFERENCES FOR CHAPTER 10

1. W.C. Gough, "Why Fusion?" U.S. Atomic Energy Commission, 1970, U.S. Government Printing Office.
2. W.C. Gough and B.J. Eastlund, "Prospects of Fusion Power," *Scientific American*, February, 1971, pp. 50–64.
3. L.M. Lidsky, "The Quest for Fusion Power," *Technology Review*, January, 1972, pp. 10–21.
4. D.J. Rose, "Controlled Nuclear Fusion: Status and Outlook," *Science*, May 21, 1971, pp. 797–808.

5. S. Glasstone, *Sourcebook on Atomic Energy*, Third Edition, Van Nostrand Reinhold Company, New York, 1967.
6. M.J. Lubin and A.P. Fraas, "Fusion by Laser," *Scientific American*, June, 1971, pp. 21–33.
7. M. Kenward, "Fusion Power Politics in the U.S.," *New Scientist*, May 18, 1972, pp. 380–382.
8. S.H. Schoor, *Energy Research Needs*, Resources for the Future, Inc., National Technical Information Service, Washington, D.C., October, 1971.
9. H. Hurwitz, Jr., "Comments on the Prospects of Fusion Power," General Electric Company 71–C–217, July, 1971.

GENERAL READING FOR CHAPTER 10

1. A.S. Bishop, *Project Sherwood—The U.S. Program in Controlled Fusion*, Addison–Wesley Publishing Co., Inc., Reading, Mass. 1958.

This is an interesting historical account of the early years of fusion research. It also acts as a good introduction to many of the basic problems which still plague development in this field.
2. W.C. Gough, "Why Fusion?" U.S. Atomic Energy Commission, 1970, U.S. Government Printing Office.

This is an interesting short review of the general problems of energy need which provide the impetus for fusion research.

PROBLEMS FOR CHAPTER 10

General Problems

10.1. Explain briefly why the D–T reaction is likely to be developed before the D–D reaction.

10.2. What is the number of electrons, protons, and neutrons in one atom of each of the following elements? $_1H^1$, $_1D^2$, $_1T^3$, $_3Li^6$?

10.3. According to the Lawson criterion, how much confinement time is necessary in the D–T reaction if the particle concentration is 100,000 billion nuclei per cubic centimeter?

10.4. Explain why it may be possible under certain circumstances to place your hand in a very hot gas without being burned.

Advanced Study Problems

10.5. In 1961 the United States committed itself to a crash program to "put a man on the moon." How does this problem differ from the fusion problem? (For instance, it is extremely unlikely that we could solve our energy problems in 10 years with fusion power.)

10.6. Using references above, and more recent documents, list and categorize according to form of operation all of the major proposals of fusion reaction forms. Show estimates where available of the effectiveness of each form.

10.7. It has been recommended by some that we should bypass breeder reaction and proceed to a major effort to develop the fusion reactor. Discuss this recommendation in detail, considering such factors as economic and environmental risks, time scales, etc.

10.8. Read Reference 1. Discuss the implications of the "fusion torch" for man's way of life.

11

MISCELLANEOUS
GENERATION TECHNIQUES

It is our will that thus enchains us
to permitted ill
We might be otherwise.

We might be all
We dream of—happy, high, majestical.

Percy Bysshe Shelley
Julian and Maddalo

We have not exhausted our study of all the ways by which man can generate electric energy, or by which he may dream to do so. In previous chapters we discussed seven ways of generating electric energy. Included were the most important present techniques and some which may never be important, at least on a large scale.

There are many approaches which we have not discussed. We cannot hope to do full justice to all of these. And yet some deserve

mention because they are important or because they are interesting. So we have gathered together in this chapter, for brief discussion, eight more generation schemes which are listed below.
Some are operating today, some are under serious study, and some
are only dreams.

A. Operating systems
 1. Gas turbines
 2. Diesel engines
B. Systems under serious consideration or development
 3. Magnetohydrodynamics (MHD)
 4. Fuel cells
C. Speculative schemes
 5. Thermal particle devices
 6. Thermal sea power
 7. Wind machines
 8. Wave machines

11.1 GAS TURBINES

A gas turbine is an engine which operates much like a steam
turbine except that the medium which flows past the turbine
blades, causing them to turn, is the gaseous product of a combustion process. A *simple cycle* gas turbine is shown in Figure
11-1. The turbine drives both the electric generator and also a
compressor whose function is to compress input air to a relatively
high pressure before it is mixed with gas or liquid fuel in the combustion chamber. The exhaust gases are released to the air after
passing through the turbine.

The major advantage of the gas turbine is its inherent simplicity.
It has no boiler or steam supply system, no condenser, and no
waste–heat disposal system. The result is that capital costs are
low: about 100 to 125 $/KW. However, the efficiency is lower
than that of modern steam plants, so operating (fuel) costs tend to
be high. Gas turbines can be started quite rapidly, with synchro-
nization and power delivery in five to 20 minutes. The low capital
cost, high fuel cost, and short starting times suggest that the pri-

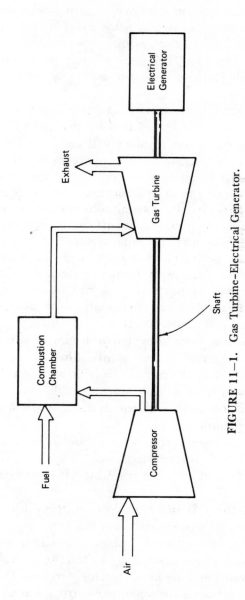

FIGURE 11–1. Gas Turbine–Electrical Generator.

mary uses for gas turbines will be for peak–power and standby sources. Recently, however, some utilities have found it desirable or even necessary to use gas turbines for an increasing number of hours each year. This is largely a result of the lack of availability, and increasing costs, of other base–load and intermediate–load plants.

Recent years have seen rapid growth of gas turbine use in the U.S. It has been estimated that by the end of 1973 the total installed capacity of gas turbines will be about 28,000 MW.(1) This represents a very significant seven percent of total U.S. capacity.

A number of factors have led to a major new interest in gas turbines. One of the most important of these is the growing shortage of capacity, resulting from construction delays, delays due to environmental restrictions, insufficient planning, and failures of other units. Gas turbine units as small as five MW or as large as 200 MW can be installed in six to eight months. Many are being installed on an emergency basis to meet critical demand needs. In addition, a wide variety of fuels can be used in gas turbines, including natural gas, process gas, distillate oil, residual oil, crude oil, naphtha, and kerosene.(2)

Gas turbines have some environmental advantages compared with other thermal power plants. Since they do not employ a steam cycle, they do not cause heat addition to water. Exhaust heat is vented from a short stack into the air. They are relatively small plants, so they tend not to disfigure their sites, and they require little ground space.

Simple cycle gas turbines have an efficiency of about 20 to 30 percent.(3) In a regenerative cycle (Figure 11–3) the compressed gas is pre–heated and the resulting efficiency is close to 36 percent.

Finally, there is much interest at this time in combining the best features of gas turbines and steam power plants, producing a so–called "combined cycle" plant. The combined cycle plant uses the hot exhaust gas from a gas turbine to provide heat to a boiler for a conventional steam generator–turbine. The gas turbine and the steam turbine drive separate electric generators. The efficiency of this device is about 40 percent. One manufacturer anticipates that combined cycle efficiencies will reach 45 to 50 percent by 1980.(3)

FIGURE 11–2. Gas turbine. One of four 156 MW floating General Electric Gas Turbine Power Blocks under way from Newport News, Virginia, to Consolidated Edison Mooring in New York City. Standard units such as this can be delivered to a customer in as little as one to six months, depending on availability of a required package. Land–based plants can be installed in six to eight months depending on size. These short installation times are an important factor in the attractiveness of gas turbines in meeting demand quickly. (*Courtesy The General Electric Go., Gas Turbine Products Division.*)

11.2 DIESEL ENGINES

Diesel engine generator units have been built for many years in the range of two to six MW. They are more efficient than gas turbines, and can be started and brought on–line faster than any unit except hydroelectric generators. Capital costs, at $100/KW, are excellent. Diesel units are particularly attractive for very small

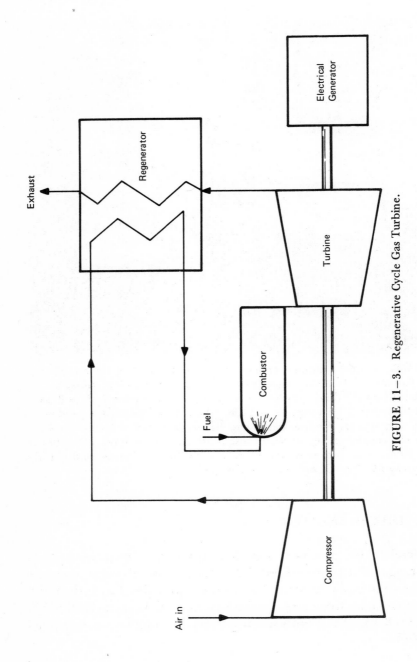

FIGURE 11–3. Regenerative Cycle Gas Turbine.

234

loads and for emergency standby systems. The major drawback of the diesel unit is its small size. The total capacity of diesel generators in 1970 was about 4,000 MW, or just over one percent of the total U.S. power capacity.

11.3 MAGNETOHYDRODYNAMICS

A magnetohydrodynamic (MHD) generator uses an ionized gas or *plasma* moving through a magnetic field to produce an electric potential. Figure 11-4 shows the basic physical action of the generator. An ionized gas is produced in a burner to the left of the figure and is blown past a magnetic field, perpendicular to the ion flow. The field deflects the positive ions to one metallic collecting plate and the negative ions to the other metallic collecting plate. The principle is simple. The application of the principle in a large efficient generator, however, has so far eluded researchers. There are a number of possible forms of a working MHD generator. The MHD unit can be developed alone or combined with a gas turbine, or with a conventional steam generator.

Figure 11-5 shows one possible complex generator using an MHD cycle and a steam cycle. Fuel is introduced to the burner along with a *seed* material such as Potassium. The purpose of the seed is to increase the conductivity of the gas sufficiently to permit practical operation of the device. The magnet deflects some of the ions to the plates, which become charged, producing a DC electric potential. This voltage can be used in this form or changed to AC. The exhaust gas passes first through an air heater which heats outside air which has been compressed by the compressor attached to the steam turbine. The heated air is then used in the burner. The hot exhaust gases from the MHD generator then pass into the steam generator, where they produce steam to drive a conventional steam turbine. The steam turbine drives both the compressor and an electric generator. The exhaust gases continue on through a seed recovery stage where the seed is captured and fed back to the burner. Since the recovery is not perfect, some make-up seed must be added. Next, the exhaust gas passes through a Nitrogen and Sulfur removal stage before being released by the stack.

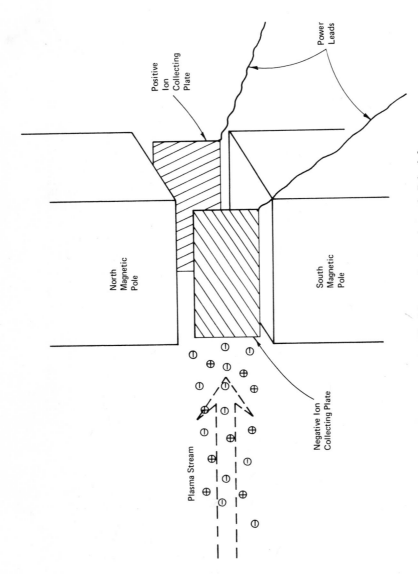

Power
Leads

Positive
Ion
Collecting
Plate

North
Magnetic
Pole

South
Magnetic
Pole

Negative Ion
Collecting Plate

Plasma Stream

FIGURE 11—4. The Magnetohydro-dynamic Principle.

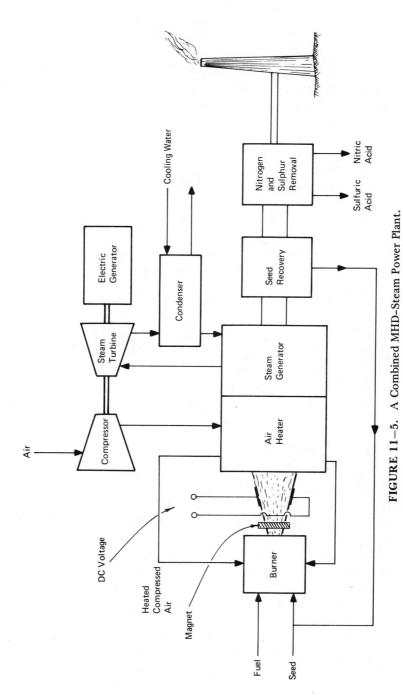

FIGURE 11–5. A Combined MHD–Steam Power Plant.

Air

Compressor

Electric Generator

Steam Turbine

Condenser

Cooling Water

DC Voltage

Heated Compressed Air

Magnet

Air Heater

Steam Generator

Seed Recovery

Nitrogen and Sulphur Removal

Nitric Acid

Sulfuric Acid

Burner

Fuel

Seed

The primary advantages of the MHD unit are its simplicity—it has no moving parts—and its high operating temperature, which leads to a high overall efficiency of perhaps 50 to 60 percent. This will provide major savings in fuel utilization and in condenser cooling.

Many problems must be solved before practical systems are developed. The burner must be designed to burn an abundant fuel, such as coal, which is dirty. To date acceptable high–volume burners operating under acceptable conditions do not exist. The very high temperatures out of the burner present new technological challenges in the design of gas channels, electrodes, and air heaters. Magnets of acceptable size and strength must be developed. As we saw in Chapter 5, high combustion temperatures lead to the production of Oxides of Nitrogen which must be removed. Adequate systems to remove Sulfur and Nitrogen Oxides are still under study.

Other MHD cycles, including a liquid instead of gaseous medium, are also being considered. Some small experimental MHD systems are under development.(4, 5, 6) It remains to be seen whether this approach can be developed well enough to make an important contribution to the generation of electric energy.

11.4 FUEL CELLS

Fuel cells are somewhat like batteries in physical form. They have two electrodes, called the *anode* and the *cathode*, and an acid *electrolyte*. Unlike batteries, however, they have a continuous fuel input. One possible fuel is Hydrogen, which is quite expensive. If the fuel cell is to be successful commercially, it will have to be able to use an inexpensive and abundant fuel such as coal or natural gas. Fuel cells have been used in space but the cost has been very great. Research is under way to reduce the cost of fuel cells substantially.(1)

Fuel cells appear to have an important potential as power supplies in remote locations, supplying small loads to single homes or apartments or small industries. In addition, there has been some recent study of the possibility of using fuel cell systems in small (perhaps 20 MW) central power plants.(7)

Fuel cells should be quite attractive environmentally, compared with many existing systems.

11.5 THERMAL PARTICLE DEVICES

Related to each other in a sense are a trio of devices which make use of heated particles to generate electric energy directly.

The electrogasdynamic generator uses a moving neutral gas, with its kinetic energy, to force charged particles to move against an electric field thus increasing the potential of the charged particles. This increased potential is then tapped off as from a battery.

The thermionic generator is an extremely simple device in which one electrode is heated enough so that electrons boil off it and flow to a nearby second electrode. A potential then exists between the two electrodes, which can be used as a source of DC electric energy.(8)

The thermoelectric generator makes use of the long-known Seebeck principle that a potential exists between two joined dissimilar conductors which are heated.

All three of these approaches are being studied but none is seriously expected to be developed in any but the distant future for central station use. The latter two may perhaps see applications in space.

11.6 THERMAL SEA POWER

We saw earlier in this book that work can theoretically be accomplished when two bodies are available at different temperatures. The greater the temperature difference, the greater the efficiency of heat–work conversion, following the Carnot efficiency relation. The oceans provide vast amounts of water at differing temperatures as a result of the sun's heat. At locations such as the Caribbean Sea, tropic waters meet and flow over arctic waters, with a temperature difference of perhaps 35° to 40° F over a separation of 1,500 to 2,500 feet.

Example 11-1

Suppose that sea water layers are available at 45° F and 75° F. What is the Carnot efficiency?

From Chapter 4, the Carnot efficiency is $e = \dfrac{T_1 - T_2}{T_1}$ where the temperatures are in °K.

$$e = \frac{535 - 505}{535} = 5.6\%$$

Because of practical problems such as losses and the energy required to lift the water, the actual efficiency would be closer to two percent to three percent.

This low efficiency should not dissuade the developer. The warm and cold waters are not fuels for which he must pay. He does, however, have to pay for the equipment required to run the plant. This can be quite significant because of the necessity of moving a great deal of water to generate a reasonable amount of energy.

A number of schemes for applying the principle exist, and some pilot plants have been developed. Some have failed because of corrosion, equipment breakdown, and other reasons. Like the tides, interest in thermal sea power ebbs and flows.(9, 10) Perhaps someday we may harness this immense source of energy.

11.7 WIND MACHINES

The wind's energy has been used to do work for man since before recorded history began. Windmills have ground grain for centuries. It is not surprising that much thought would be given to using the winds to generate electric energy.(11) Two major factors have inhibited the development of wind machines for electric energy generation. First, the winds are greatly variable. They even compare poorly with a flowing stream as a source of continuous and dependable energy. Second, a very large machine or many machines together are required to provide sufficient energy to justify most proposals. Wind machines so far are therefore best suited to some intermittent tasks such as the pumping of water from a well.

Wind machines can be used profitably by combining the intermittent input energy with a storage system which can be tapped at some later time to provide dependable electric energy. (See Figure 11-6.) One such storage system is a simple battery. This scheme is used in Sweden to supply electricity to remote automatic telephone exchanges. The system includes sensing devices to control the charge delivered to the battery so that it is not overcharged. Another possible storage scheme is a water storage or pumped-storage hydro system. The major problem in this case would no doubt be obtaining enough energy to justify the installation costs. Once again, "free" energy may just be too expensive.

11.8 WAVE MACHINES

Both the winds and the waves contain extraordinary quantities of kinetic energy. The energy of the waves is perhaps even more dramatic, since it produces the rolling, crashing breakers. It is certainly not surprising that man would try to harness this source, as he has. In 1909 the California Wave Motor Co. built Alva Reynold's "Ideal Wave Motor," a contraption using panels moving under water to transfer the energy of the waves to an electric generator. The device worked! Improvements were promised, and the device was to revolutionize the power industry. But the improvements didn't work out. For years the device lighted lamps on the wharf at Huntington Beach, until one day a storm washed away the machine. The power industry looked elsewhere for its revolution.

Interest in wave machines did not die with that storm, but neither has it led to major new triumphs. Experiments continue now and again along the coasts. It is difficult to believe, however, that large sections of ocean shores would ever be dedicated to this form of electric energy generation, whether it is financially feasible or not.

FIGURE 11–6. An experimental wind machine. The energy of the winds has been used for centuries in wind machines. Today a number of fairly small wind machines generate electric energy, usually to pump water or charge a storage battery. Experiments continue to seek better energy conversion, primarily for remote low–energy sites. The device above is an experimental 500 watt machine which has been tested at Oklahoma State University in Stillwater, Oklahoma. (*Courtesy School of Electrical Engineering, Oklahoma State University.*)

11.9 CONCLUSION

There are many more ways in which the energy of nature might be harnessed to do man's work. It seems unlikely that the search for a new Ideal Machine will ever cease. Some will be tried and will fail. Some will be limited successes for special applications. And some may truly revolutionize the electric power industry. But none will be free. Two questions will always demand answers:

Can it be done?
Is it worth the price?

REFERENCES FOR CHAPTER 11

1. E.W. Springer, "Gas Turbine Options and Economics," *Combustion*, April, 1973, pp. 36–39.
2. A.D. Foster, "Gas Turbine Fuels," *Combustion*, April, 1973, pp. 4–14.
3. N.R. Dibelius and E.W. Zeltman, "Gas Turbine Environmental Impact Using Natural Gas and Distillate Fuel," Report No. 73-GTD-6, General Electric Company, Schenectady, N.Y., February, 1973.
4. M. Petrick, "Liquid–metal Magnetohydrodynamics," *IEEE Spectrum*, March, 1965, pp. 137–151.
5. D. Bienstock *et. al.*, "Environmental Aspects of MHD Power Generation," Proceedings of the 1971 Intersociety Energy Conversion Engineering Conference, Boston, Mass., August 3–5, 1971.
6. A.M. Bueche, "An Appraisal of MHD–1969," Office of Science and Technology, Washington, D.C., February 3, 1969.
7. W.J. Lueckel *et. al.*, "Fuel Cells for Dispersed Power Generation," Paper No. T72-235-5, *Proceedings of the IEEE Winter Meeting*, New York, Jan. 30–Feb. 4, 1972.
8. V.C. Wilson, "Thermionic Power Generation," *IEEE Spectrum*, May, 1964, pp. 75–83.
9. S. Walters, "Power in the Year 2001, Part 2—Thermal Sea Power," *Mechanical Engineering*, October, 1971, pp. 21-25.

10. H. Ting, "A Possible Breakthrough of Exploiting Thermal Power from the Sea," *Combustion*, August, 1970, pp. 16–19.

11. J. McCaull, "Windmills," *Environment*, Vol. 15, No. 1, Jan./ Feb. 1973, pp. 6–17.

GENERAL READING FOR CHAPTER 11

1. *The 1970 National Power Survey—Part 1,* Federal Power Commission, U.S. Government Printing Office, December, 1971.

   Chapters 8 and 9 review a number of the schemes mentioned in this chapter. The material is descriptive and easily understood.

2. S.W. Angrist, *Direct Energy Conversion*, Allyn and Bacon, Inc., Boston, 1971.

   This textbook provides much technical detail in areas such as thermoelectric, photovoltaic, thermionic, magnetohydrodynamic and fuel cell generators.

3. B. Chalmers, *Energy*, Academic Press, New York, 1963.

   This text, less technical than the previous listing, discusses many of the topics of this chapter.

4. H.C. Hottel and J.B. Howard, *New Energy Technology—Some Facts and Assessments*, M.I.T. Press, Cambridge, 1971.

   This book has some fairly technical details on recent work in gas turbine and magnetohydrodynamics technology.

PROBLEMS FOR CHAPTER 11

Advanced Study Problems

11.1. Find in a recent article or book a newly proposed scheme for generating electric energy. Discuss its basic physical basis, its energy source, and the problems associated with its development. Also, estimate its prospects for significant development.

11.2. Compare in detail the relative air pollution produced by gas turbines and other fossil–fueled power plants. Evaluate the prospects for reducing air pollution from these systems in the future.

11.3. Discuss in detail the problem of evaluating fuel cells as a prospective source of electric energy at remote sites with small energy demands. What other schemes would be competitive in such a market? Consider the economic evaluation of competing schemes as well as you can.

11.4. Find additional literature on wave machines and write a brief history of this approach to electric energy generation. What factors have inhibited the growth of this form of power?

11.5. Repeat 11.4 for wind machines.

12

WASTE HEAT

There can be no economy where there is no efficiency.

Disraeli

We saw in Chapters 5 and 6 that huge amounts of energy are lost as waste heat in fossil–fuel and nuclear power plants. In this chapter we restrict our attention to the heat given up in the condenser and ignore other losses.

We start with some calculations of the amount of heat wasted. This leads us to a consideration of the effect which added heat has on the biological world. Next we consider the ways in which heat can be dissipated, finishing with a study of the economic and environmental considerations relevant to each of the cooling schemes.

12.1 CONDENSER COOLING REQUIREMENTS

We want to know how much heat must be carried off by condenser cooling water. This will give us a sense of the magnitude of
the problem before we ask what effect this heat has, and then what
we might do about it.

Burning fuel in a thermal power plant (either fossil or nuclear)
produces a certain amount of heat energy, E_f. A fraction η of this
energy is available as electric energy E_g ($= \eta E_f$). A fraction $(1-\eta)$ of
E_f is lost as waste heat. In a fossil–fuel plant about 85 percent of
this waste heat is carried off by the condenser cooling water, and
15 percent is lost in stack gases and other miscellaneous forms. In
nuclear plants about 95 percent of the waste heat is carried off by
the cooling water. Hence the total energy dissipated by the condenser is:

$$E_c = r(1 - \eta) \, E_f \qquad\qquad\qquad (12\text{-}1)$$

where:

$$r \cong \begin{cases} 0.85 \text{ for fossil–fuel plants} \\ 0.95 \text{ for nuclear plants} \end{cases}$$

In the discussion to follow we shall be concerned with the *time
rate* at which heat energy is: (a) produced; (b) transformed into
electric energy, and (c) lost as waste heat.

Working with the rate of heat flow is convenient because it
permits a direct relation to the electric *power* generated by a plant.
We shall represent heat rates by $\dot{E}$ where the dot stands for a time
rate. Also we shall commonly refer to $\dot{E}_g$ as P_g, the generator
power. (Recall that in Chapter 1 we defined power as the time
rate at which energy is made available or work is done.) Using
these conventions we write:

$$P_g = \dot{E}_g = \eta \dot{E}_f \qquad\qquad\qquad (12\text{-}2)$$

$$\dot{E}_c = r(1 - \eta) \, \dot{E}_f \qquad\qquad\qquad (12\text{-}3)$$

where:

$\dot{E}_f$ = rate of heat production by the fuel

$\dot{E}_g$ = generator power

$\dot{E}_c$ = rate at which the condenser cooling water carries off heat

η = efficiency of the plant

r = fraction of waste heat carried off by the condenser

The units of these heat rates are KW. The relationships above are summarized in Figure 12-1.

Next, we seek a direct relation between the condenser waste heat rate and the generated electric power. This relation is available by solving Equations (12-2) and (12-3) to eliminate $\dot{E}_f$. This yields:

$$\dot{E}_c = r \frac{1 - \eta}{\eta} P_g \tag{12-4}$$

Example 12-1

Find the condenser cooling water waste heat rate for a modern 1,000 MW fossil-plant with an efficiency of 40 percent. Compare this with the power dissipated by the same size (same P_g) nuclear plant with an efficiency of 33 percent.

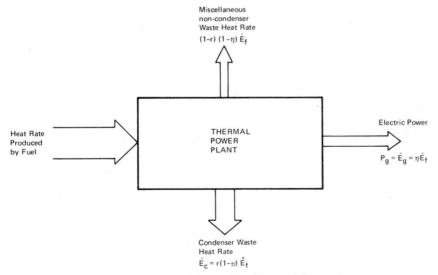

FIGURE 12—1. Heat Rates in a Thermal Power Plant.

Fossil: $\dot{E}_c = 1{,}000 \times \dfrac{1 - .4}{.4} \times 0.85 = 1{,}275$ MW

Nuclear: $\dot{E}_c = 1{,}000 \times \dfrac{1 - .33}{.33} \times 0.95 = 1{,}930$ MW

The very interesting and important result shown in Example 12–1 is that, for these assumed efficiencies which are reasonable modern plant numbers, *nuclear plants must dissipate about 50 percent more condenser waste heat than fossil plants.*

Now that we have the condenser waste heat rate, in Equation 12–4, we shall want to find out what the flow rate of condenser cooling water must be. With the equivalence relation 3,413 BTU = 1 KWH, we can write (12–4) as:

$$\dot{E}_c = 3{,}413 \, \frac{r(1 - \eta)}{\eta} \, P_g \left(\frac{\text{BTU}}{\text{Hour}}\right) \tag{12–5}$$

where P_g is expressed in KW.

Example 12–2

What is the answer to the problem in Example 12–1 expressed in BTU/Hr.?

Fossil: $\dot{E}_c = 4.36 \times 10^9$ BTU/Hr.

Nuclear: $\dot{E}_c = 6.59 \times 10^9$ BTU/Hr.

Before we can proceed further we need to know how many degrees we can or should increase the cooling water temperature. The amount of cooling-water temperature increase depends on the construction of the condenser and the rate at which water passes through the condenser. For technical reasons too complex to discuss here, the temperature increase should not be less than about 10° F nor more than about 20–25° F. A typical change is 15° F. We shall call this change ΔT.

Recall that one BTU raises the temperature of one pound of water one degree F. Hence if we divide $\dot{E}_c$ in Equation 12–5 by ΔT we have the required water–flow rate in lbs./hr. To convert this to cubic feet per second we divide by

$$3,600\left(\frac{sec.}{hr.}\right) \times 62.4\left(\frac{lbs.}{cu.\ ft.}\right)$$

Introducing these three factors into Equation 12–5 yields:

$$Q = \frac{0.0152r\ (1 - \eta)\ P_g}{\eta \Delta T}\ (cfs) \qquad\qquad (12\text{--}6)$$

Example 12-3

Determine the required flow rate of cooling water in a 1,000 MW nuclear power plant with an efficiency of 33 percent. Assume a condenser temperature change $\Delta T = 15°$.

$$Q = \frac{0.0152 \times 0.95\ (0.67) \times 10^6}{0.33 \times 15}$$

$$= 1,950\ cfs$$

Example 12-4

Calculate the cooling water in acre-feet per year required by the plant described in Example 10-4.

$$Q = 1,950 \times (3,600)\ (8,760) = 6.15 \times 10^{10}\ \frac{cu.\ ft.}{year}$$

$$= \frac{6.15 \times 10^{10}}{43,560} = 1.41 \times 10^6\ \frac{ac.\ ft.}{year}$$

It is clear from the above examples that thermal power plants require very large water flow rates for cooling. In the past it has been common to use river, ocean, or lake water run through the condensers once and then returned to the source. This is called *once–through cooling*. In recent years, however, there has been increasing concern about the effect of heated water on the biological environment. This concern has led to the development of a number of alternatives to once–through cooling. Before looking into these we shall want to ask what effect added heat has on the biological residents of the cooling–water source.

12.2 BIOLOGICAL EFFECTS OF ADDED HEAT

To the conservationist, addition of heat to water often becomes "thermal pollution." To the ardent proponent of power development it has been "thermal enhancement" or even "thermal enrichment," on occasion. We shall adopt here the more non-committal term *thermal addition* for the general case of added heat. Whether added heat enriches or pollutes is very much a matter of circumstances, including which biological entities are under study. It seems reasonable to state that thermal pollution has occurred when added heat produces effects which are detrimental to other uses of the water. The problem then becomes one of determining whether the heat is detrimental in some way.

The effect of added heat on water has been studied for many years. Recently this study has intensified, but the fact remains that there is very much we do not yet know. (See Reference 1, pp. 137–144.) This will no doubt be an area of extensive research in the years ahead. In the meantime we have no clear and unambiguous answer to the question, "Does added heat cause unacceptable levels of thermal pollution?"

Added heat has a number of complex effects on the water itself. In general as water is heated there is a tendency toward:

1) Decreased oxygen capacity
2) Increase in chemical reaction rate
3) Stratification, or formation of layers of water at different temperatures
4) Increased viscosity

Heat addition affects aquatic life in many ways, some direct and apparent, some far less obvious. The life of both plants and animals is affected. Plant life, which uses sunlight and nonliving components of the water for nutrition, becomes food for aquatic animal life. But temperature affects the life of important plants such as algae. As temperature is changed some forms become more dominant, some decline. The result can be a major change in the ecology of the source of the cooling water. Plants are only a part of the complex aquatic ecosystem. Other important members of the system are bacteria and fungi, phytoplankton, zooplankton, and fish. All have varying tolerance to temperature change.

To consider the effects of temperature in more detail we center our attention on fish, usually the most apparent residents of the water source. Heat addition can affect fish in a number of ways.

First, death of fish can be directly related to heat addition. Fish have maximum allowable *temperature* limits, and maximum allowable *temperature change* limits. Laboratory studies have given us some answers to the questions of temperature tolerance of certain fishes. Table 12-1 summarizes a part of one such study.(2) The water temperatures given are for survival of 50 percent of the test fish over 24 to 133 hours. Results of such tests give no final answers to temperature limits for fish, but they do provide a start in this highly important study. Outside of the laboratory we have some data on fish kill in reported incidents. One study lists 18 kills reported to the Federal Water Pollution Control Administration between 1962 and 1969.(3) The total number of fish killed exceeded 700,000, with a worst-case kill of over 300,000 fish at Sandusky, Ohio in 1967.

Second, increased heat may interfere with spawning. Spawning is affected by temperature in two ways. First, fish depend on temperature as a signal for migration and spawning. Temperature is a kind of seasonal yardstick. Second, eggs will not hatch above a certain temperature, depending upon the species of fish.

TABLE 12-1.

Minimum and Maximum Temperatures for Certain Freshwater Fishes

Species	Acclimated to °F	Minimum Temp. °F	Hours	Maximum Temp. °F	Hours
Largemouth Bass	68.0	41.0	24	89.6	72
	86.0	51.8	24	93.2	72
Bluegill	59.0	37.4	24	87.8	60
	86.0	51.8	24	93.2	60
Channel Catfish	59.0	32.0	24	86.0	24
	77.0	42.8	24	93.2	24
Brook Trout	37.4	—	—	73.4	133
	68.0	—	—	77.0	133

Third, fish suffer internal biological effects from added heat. Changes in temperature cause changes in a number of the fish's bodily functions, such as respiration, food intake, activity level, growth rate, and lifespan.(4) Occasionally this effect may, in some cases, be used for good.

Fourth, fish may be killed through indirect effects. As we saw above, fish are only a part of a complex aquatic ecosystem. Changes in that system may lead to the death of the fish. Particularly important are factors such as reduced oxygen, changes in food supply, an increase in toxic substances, or a decrease in resistance to such substances, and changes in the stratification and light penetration of the body of water.

Finally, competitive replacement by other species may occur. Since some fish are more active or more adaptable to a given temperature, a change in temperature may change the dominant species. Cold water species such as trout may be forced out of a system as warm water species begin to dominate.

There is much more which can be said about harmful thermal effects on aquatic life.(5, 6) We turn our attention now to some beneficial aspects of heat addition to water.

The fact that aquatic life changes its functional characteristics can in some cases be used to advantage. Major interest exists today in accelerating growth of shellfish and some fish species by exposing them to a temperature at which growth is most rapid. This permits more rapid harvesting of edible fishes.

Most of the effects of temperature elevation on aquatic life are presently considered to be undesirable. In addition, there are a number of areas in the country in which sufficient water is not available for once-through cooling. These two factors have led in recent years to significant new efforts toward finding alternative ways to cool condenser water. In the next section we discuss once-through cooling and some of its major alternatives.

12.3 HEAT DISSIPATION TECHNIQUES

There are a number of available schemes for cooling condenser water. The four general classes are:

1) Once-through cooling
2) Cooling ponds
3) Cooling towers
4) Heat transfer to other systems

There are a number of ways of operating each of these so that the total number of possible schemes is well over 20. We shall avoid most of the variations on each scheme and discuss only the most common forms of each.

In all schemes it is necessary to dissipate or spread out the heat. Eventually the heat is diffused over a sufficiently large part of the earth so that its effect is not apparent. Heat does not accumulate on the earth; it radiates out into space.* The difficulty is not with the amount of heat added, but rather with its concentration. Hence each proposed scheme must provide a mechanism for dispersing this concentrated heat with as few adverse side effects as possible.

In the once-through cooling scheme, water withdrawn from a river, lake, bay or ocean is run through the condenser with a temperature increase of perhaps 10° to 25° F, and is then returned to the source. This scheme is sketched in Figure 12-2.

A number of factors must be considered in designing a once-through cooling system. The water must be sufficient to meet the required flow rates as given by Equation 12-6. If the source of cooling water, sometimes called a *heat sink*, is a river, the minimum river flow rate should be somewhat larger than Q in the equation. The greater the flow rate of the river, the greater is its potential to disperse or spread the heat quickly as the heat moves downstream from the plant. If the heat sink is the ocean, the intake and discharge tubes must be sufficiently far apart that no significant part of the heat discharged is present at the intake tube. This effect will depend on such factors as tides and ocean currents.

Once-through cooling is used very extensively in the United States at ocean, river, lake and bay sites. Opposition to this form of cooling has arisen at all these types of sites. Although the ocean

---

*There is a balance between heat received from the sun and heat radiated into space. This balance is not affected by local additions of heat.

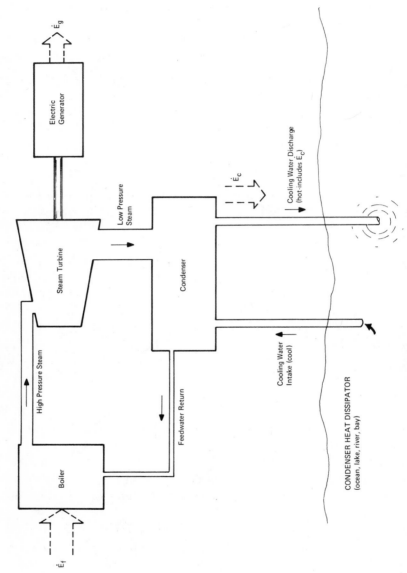

FIGURE 12–2. Once-through Cooling.

256

is potentially an almost infinite source of mixing and dispersing heat, the effects near the discharge point may be harmful or disastrous to local marine life, including kelp beds, fish populations, and other forms of ocean life. Essentially the same problem limits the use of rivers and lakes, except that here, if the body of water is small, a large part if not all of the body may increase in temperature.

The possible damage to aquatic life is one of the factors which has led to the search for an alternative to once-through cooling. The second factor is that it is not always convenient or desirable to place a plant near a large natural source of cooling water. Hence we need some other ways to cool condenser water.

The first obvious alternative is to build our own small lake or *cooling pond*. There is then no question of disturbing an existing natural body of water. In fact it may even be possible to develop a desirable new water facility, at higher than normal temperatures, suitable for water sports and warm water fish. The disadvantage of the pond is that its methods of getting rid of its heat (principally evaporation and radiation) are fairly slow, so that relatively large ponds are required. A common rule of thumb is about one to two acres per megawatt of installed capacity. That means about 1,000–2,000 acres (two to three square miles) for a large modern 1,000 MW plant. This is a rather large area, which may not be available at many sites.

One technique for decreasing the amount of required pond area is to build a mechanism into the pond which sprays water into the air and increases the water area available for evaporation. This forms a so-called *spray pond*.

If a large natural body of water is not available and a pond is not desirable, we try to find a system which does what a pond does but in less space. One such system is a *cooling tower*.(7, 8) We limit our discussion here primarily to *wet cooling towers*. In the wet cooling tower the heated condenser water is delivered to an open-top tower, pumped up 20 to 100 feet, and allowed to "rain" down to the floor of the tower as air moves upward through the tower. This greatly increases the surface exposure of water to air and increases the evaporation (and a little bit of radiation)

FIGURE 12—3. A hybrid of cooling techniques. For years the Sacramento River was used for once–through cooling of the Pittsburgh Plant in California. When a new 790 MW unit (stack on the left) was added in 1972, water pollution control laws prohibited further use of the river for cooling. Pacific Gas and Electric Company built a long U–shaped spray pond (above and to the left of the stacks) for the new unit. Note also the oil off–loading pier and the oil storage tanks. (*Courtesy Pacific Gas and Electric Co.*)

enough to dissipate the required heat. The principal byproduct of the cooling tower is water vapor. This system is sketched in Figure 12-4.

There are two important types of wet cooling towers. Both are shown in Figure 12-5. One is a forced or induced draft tower in which air motion is caused by a mechanical fan placed either on top of the tower (induced) or on the bottom sides (forced). The second is the natural draft tower. It requires sufficient height— about 400 feet—that a flow of air from bottom to top occurs naturally. Both types are used in the United States. The latter type is becoming increasingly popular. The forced or induced draft

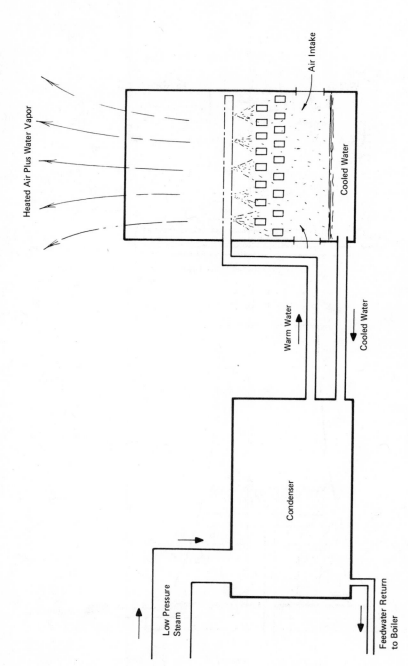

FIGURE 12–4. Cooling Tower System.

259

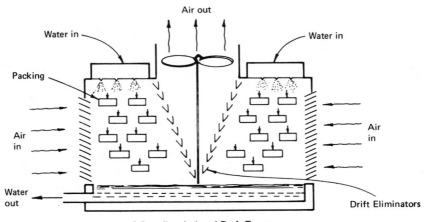

a) Crossflow Induced Draft Tower

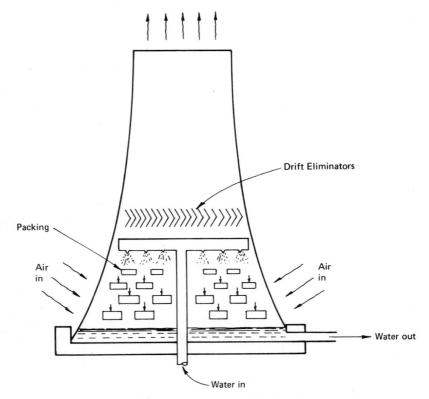

b) Hyperbolic Natural Draft Wet Tower

FIGURE 12—5. Wet Cooling Towers.

tower is relatively small (75-100 feet high), and is made of wood. The natural draft tower is a massive (usually concrete) structure perhaps 400 feet high and nearly as wide. These brief descriptions give all of the basic ingredients for analyzing the advantages and disadvantages of each.

Aesthetically, the smaller tower is almost certainly preferable because of its low profile. The natural draft tower has a fundamentally pleasing hyperbolic shape but its massiveness usually makes it far out of proportion to its surroundings. In terms of capital costs, the natural draft tower is much more expensive than the smaller tower. However, maintenance and operation costs (which include the cost of electric power to run the fans) are much greater for the forced or induced draft towers. The trade-offs evident here are sufficiently balanced so that it is not clear which tower is in general more desirable, and both forms are being adopted.

One of the major disadvantages of the wet cooling tower is its water loss due to evaporation. An alternative approach is to keep the cooling water inside heat-radiating pipes in the tower so that it is not exposed to the air. This saves water loss from evaporation, but it also cuts down the efficiency of the cooling system, and increases its cost significantly. This so-called *dry cooling tower* has not been developed extensively to date. However, growing water shortages in some areas and possible environmental effects of wet towers may lead to increased interest in dry towers.(9, 10)

The final method for dissipating condenser heat is one which is being viewed with a great deal of interest at this time. This is the general technique of using the waste heat for some useful purpose. It has been proposed that waste heat be used for heating offices and homes, increasing agricultural yields by using warm water for irrigation, salt-water desalination plants, sewage treatment plants, and a variety of other processes. Such proposals seem very attractive and in fact many of these things are actually being done on at least a small scale. Whether development along such lines will be extensive is not clear. It is certainly attractive and effort no doubt will be directed toward such goals.

12.4 MAKE-UP WATER

We have saved for a separate section one problem common to ponds and wet towers, because it is a difficulty which will almost certainly be increasingly serious as time goes on. At first glance the use of ponds or towers seems particularly attractive from the standpoint of water use because the same supply of water appears to cycle continually from condenser to cooler, back to condenser, etc. Unfortunately, however, there are losses and they are not, in general, insignificant. Whether the cooler is pond or tower there are three important sources of loss of water:

1) Evaporation (1–3%)
2) Drift (0.1–0.5%)
3) Blow-down (1–3%)

Evaporation is the basic mechanism for the removal of heat from the water. We can get a rough idea of the water loss with some simple calculations. Suppose that the cooling scheme must remove 20° F of temperature increase before the water is re-used. Recall again that one BTU increases the temperature of one pound of water by one degree F. Thus each pound of cooling water must lose 20 BTU. But one pound of water requires roughly 1,000 BTU to evaporate. Therefore, if all of the heat is carried off by evaporation, about one of every 50 pounds of water (or two percent) must evaporate.

Drift is the process by which small water particles (not vapor) are carried off by the air. This can usually be reduced to acceptable limits by tower design.

Blow-down is a process of flushing out and discharging the water supply in the system periodically. It is necessary because evaporation leaves behind solids which tend to concentrate in the water system. The frequency of this flushing process depends on the degree of contamination of the source of water.

It is apparent from the above that a non-trivial percentage (perhaps two to four percent) of the circulating water is continually being lost and must be replaced from a source of make-up water. The magnitude of this demand is evident from the following example. (For more detail see Reference 11.)

Example 12-5

Consider a 1,000 MW nuclear power plant with an efficiency of 33 percent and a condenser $\Delta T = 15°$ F (as in Examples 12-3 and 12-4). If the water loss is three percent, calculate the required water flow rate of make-up water. From Example 12-4 we have a condenser flow rate $Q = 1.41 \times 10^6 \frac{\text{ac.-ft.}}{\text{yr.}}$.

Make-up flow rate: $Q = 0.03 \times 1.41 \times 10^6$

$$= 42,300 \frac{\text{ac.-ft.}}{\text{yr.}}$$

$$= 58.5 \text{ cfs.}$$

In many dry regions this is a significant amount of water which may even be a crucial determining factor in a site location decision.

The loss of water inherent in wet tower systems is the primary impetus for the development of dry cooling towers. The dry tower concept is severely handicapped by the fact that its heat loss mechanisms are much less effective than evaporation. The result is that dry towers must be much larger and more expensive than wet towers.

12.5 THE ECONOMICS OF COOLING SYSTEMS

There is no general agreement about costs of alternative cooling method. These figures depend on such variables as plant size, site location (including environmental conditions), and the status of the economy. It is possible, however, to indicate some approximate cost levels which at least suggest relative costs with some accuracy. Table 12-2 is a summary of values reported in the recent literature.

Once-through cooling, where it is possible, is almost certain to be the most economical solution. Ponds and wet towers, both mechanical and natural draft, compete fairly evenly on an economic basis; certainly all three appear commonly in practice. Selection is often made on grounds other than economic.

TABLE 12-2.

Costs of Cooling Water Systems

System	Fossil ($/KW)	Nuclear ($/KW)
Once–Through	2–6	3–8
Cooling Ponds	5–8	7–12
Mechanical Draft Wet Tower	5–10	7–13
Natural Draft Wet Tower	6–14	8–20
Mechanical Draft Dry Tower	20–50	30–65

Example 12-6

Find the added cost in mills/KWH if a natural–draft wet tower is used with the fossil–fuel plant described in Table 5-2. Assume that the cost is $10 per KWH.

In this case we consider only the fixed charge (13.66 percent) on $10, and neglect the maintenance costs. Yearly fixed charges per kilowatt are:

$$0.1366 \times 10 = \$1.37$$

This represents an increase of 6.85 percent over the $20 total annual capacity cost. This of course implies a 6.85 percent increase in the fixed energy cost, which was 4.15 mills/KWH. This increase is *0.284* mills/KWH. Total energy costs were 6.17 mills/KWH. 0.284 represents a 4.6 percent increase in energy cost. Consistent with the results of this example, cooling costs in general tend to run about five to ten percent of total energy costs.

12.6 COOLING SYSTEMS AND THE ENVIRONMENT

Perhaps no power plant environmental problems are as complex or difficult to resolve as those of the cooling system. Alternatives to once-through cooling are themselves largely a response to an inadequately-understood threat to the environment. In their turn, the alternatives have an impact which is not easily assessed. In this section we describe some of the environmental concerns without in any way attempting to make a general statement about which system is "best." That question can be approached only in the particular and not in the general.

The problems of once-through cooling have been discussed to some extent earlier in this chapter. Heat added to natural waters tends to change the ecological balance. This may result in either subtle or catastrophic effects on marine life. There also may be some desirable effects, such as increased growth rates of some food fish.

Cooling ponds do not affect an existing natural body of water. However, they do have the disadvantage of taking up a fairly large amount of land (perhaps one to three square miles). Often this land is not available on the market. In other situations it may be available, but removing the land from use as land may be considered environmentally undesirable. Warm water ponds of this size may also cause significant local weather changes such as the development of fog banks during certain times of the year. Ponds can be effectively developed in some areas as multi-use facilities. Their temperatures immediately suggest warm-water fish breeding, and warm-water sports, including fishing, swimming, sailing.

Wet towers have the disadvantage of releasing large amounts of water vapor into the air, with possible local meteorological effects such as fogging, or icing of nearby roads.(12) In addition, towers release to the atmosphere other undesirable particles or gases present in the cooling water. Natural draft towers, which can be as much as 400 feet high, are considered to be aesthetically undesirable in many sites.

Dry towers do not release water vapor. They do release warm air which could have minor local effects, and they are somewhat larger than wet towers.

As we said earlier, there is no "best" solution. We are faced here with the problem of dispersing to the environment a large concentration of heat. The effect which each method has depends on many variables. Only when these are considered fully for a particular site can a choice be made. And still that choice must be made in the midst of much uncertainty.

Before closing this chapter we turn briefly to the question of costs and environmental benefits. At first it may seem that cooling towers are added to a plant only for environmental reasons. Their cost would then represent an internalizing of the cost of preventing water pollution. However, the situation is not quite that simple. Thermal power plants require some kind of cooling system to operate, quite apart from environmental concerns. Hence a certain basic portion of cooling costs must be assigned to plant operation. If a more sophisticated system is added to protect the environment, then only the incremental costs beyond the base cost should be charged to the environment. If a plant is sited in a normally dry region, for other than environmental reasons, the cost of a water-conserving cooling system likewise cannot be attributed to environmental concerns. None of this affects the final cost of energy to the consumer, but only the question of whether or not the added cost is a part of "cleaning up the environment."

REFERENCES FOR CHAPTER 12

1. National Academy of Engineering, *Engineering for the Resolution of the Energy-Environment Dilemma*, Washington, D.C., 1972.
2. Federal Power Commission, *Problems in Disposal of Waste Heat from Steam-Electric Plants*, Federal Power Commission Staff Study, 1969.
3. S.C. Bloom, *Heat—A Growing Water Pollution Problem*, in *Environment Reporter*, Monograph No. 4, The Bureau of National Affairs, May 1, 1970.
4. J.R. Clark, "Thermal Pollution and Aquatic Life," *Scientific American*, March, 1969, pp. 19–28.
5. P.A. Krenkel and F.L. Parker, *Biological Aspects of Thermal Pollution*, Vanderbilt University Press, 1969.
6. W.S. Lee *et. al.*, *Environmental Effects of Thermal Discharges*, The American Society of Mechanical Engineers, New York, 1970.
7. R.D. Woodson, "Cooling Towers," *Scientific American*, May, 1971, pp. 70–78.

8. J.B. Dickey and R.E. Cates, "Managing Waste Heat With the Water Cooling Tower," The Marley Co., Kansas City, Missouri, 1970.

9. H. Heeren and L. Holly, "Dry Cooling Eliminates Thermal Pollution-II," *Combustion*, November, 1972, pp. 17-28.

10. R.E. Holmes and S.J. Basham, Jr., "A Dry Cooling System for Steam Power Plants," *Proceedings of the Intersociety Energy Conversion Engineering Conference*, Boston, Mass., August 3-5, 1971.

11. P. Leung and R.E. Moore, "Water Consumption Determination for Steam Power Plant Cooling Towers," *Combustion*, November, 1970, pp. 14-23.

12. G.F. Bierman *et. al.*, "Characteristics, Classification and Incidence of Plumes from Large Natural Draft Cooling Towers," *Combustion*, October, 1971, pp. 25-31.

GENERAL READING FOR CHAPTER 12

1. *Cooling Towers*, American Institute of Chemical Engineers, New York, 1972.

   This is a very good set of papers on cooling-tower problems. They are technical papers but many of them can be easily read and understood by persons without technical expertise.

2. National Academy of Engineering, *Engineering for the Resolution of the Energy-Environment Dilemma*, Washington, D.C., 1972.

   Pages 103 to 150 give a very good technical and general survey of cooling problems covering all points discussed in this chapter. An extensive list of references is included.

3. D.A. Berkowitz and A.M. Squires, *Power Generation and Environmental Change*, M.I.T. Press, Cambridge, Mass., 1971.

   This book has five papers on waste heat problems in addition to groups of papers on fossil-fuel power, hydroelectric power, and nuclear power.

4. P.A. Krenkel and F.L. Parker, *Biological Aspects of Thermal Pollution*, Vanderbilt University Press, 1969.

This is an excellent set of general readings on engineering as well as biological effects of heat addition.

5. P.H. Cootner and G.O.G. Löf, *Water Demand for Steam Electric Generation*, The Johns Hopkins Press, Baltimore, 1965.

This is a very good engineering and economic analysis of cooling systems. It is appropriate only for the advanced student or practicing engineer.

PROBLEMS FOR CHAPTER 12

General Problems

12.1.   Water flow rates are clearly of great importance in cooling problems. Such flow rates may be expressed in cubic feet per second (cfs), gallons per minute (gpm) or acre–feet per year (afy). Make a table which shows how these measures are related.

12.2.   Consider a large, modern, 1,400 MW coal–fired steam plant with an efficiency of 40%. If the condenser temperature change is 20°, calculate the condenser cooling water flow rate in cfs, gpm, and afy. (Q = 1,350 cfs = 607,000 gpm = 977,000 afy)

12.3.   Repeat Problem 12.2 under the assumption that the plant is nuclear and has an efficiency of 33%. (This is the kind of first–order calculation that a siting engineer would have to make in considering plant–type and water availability.) (Q = 2,050 cfs = 921,000 gpm = 1,485,000 afy)

12.4.   For the plant described in Problem 12.2 find the make–up water flow rate in cfs if the water loss in the cooling system is 3% of the condenser water flow.

12.5.   Assume that the plant described in Problem 12.3 uses a natural draft tower with a water–loss of 2.5%. How large a lake would have to be created if make–up water were to be stored for one year's use? Assume an average lake depth of 50′ and give your answer in square miles. (1.16 sq. mi.)

12.6.   Write an equation for the miscellaneous non–condenser waste heat rate. Call it $\dot{E}_m$, a function of r, η, and P_g. Calculate $\dot{E}_m$ for Problem 12.2. (Assume r = 0.85.)

12.7. Repeat Problem 12.6 for the five cooling systems given in Table 12-2. Make a table of total energy costs and the percentage increase from each particular system.

Advanced Mathematical Problems

12.8. Write an equation for the cooling water flow expressed in gallons/KWH for a given ΔT, r, and η. Calculate this flow for $\Delta T = 15°$ F, r = 0.95, and η = 0.34. (50 gal./KWH)

12.9. Suppose that P_g varies with time and cooling water varies proportionately. Write an equation for the total number of cubic feet of water required to cool a plant with a power output of $P_g(t)$. Let t be expressed in hours.

12.10. Assume that of all the electric energy produced in the United States in 1971 80% was produced by fossil–fuel plants using once–through cooling, and having an average efficiency of 33%. What was the average flow rate in cfs? Compare this with the flow rate of the Mississippi River.

Advanced Study Problems

12.11. Write a short paper on the history of the dry cooling tower and its prospects for future use.

12.12. Write a paper on the effects of heat on one specific aquatic plant or animal.

12.13. It is often suggested that waste heat should be put to some useful work, such as heating, irrigation, etc. After doing some research on this topic, discuss situations in which such utilization may be feasible, and some of the problems associated with it.

13

ENERGY RESOURCES

A crisis exists right now. For the next three decades we will be in a race for our lives to meet our energy needs.

> John A. Carver, Jr.
> Federal Power Commission

In the late spring of 1973, 50 influential energy experts, representing a broad spectrum of industry, government and conservation groups, met in California to discuss that state's energy problems. The result was an astounding set of proposals that could change the life styles of most Californians. If such measures are necessary in California they are probably also necessary in the entire country. Some of the proposals were:

1) A statewide 50 mile per hour speed limit for all motor vehicles
2) A very steep "weight tax" on virtually all cars that could

add $2,000 or more to the tax on a standard-size American car

3) Diversion of highway funds into rapid transit
4) A ban on inefficient air conditioners and new electric heating installations
5) Reopening of off-shore oil well drilling
6) Large-scale expansion of nuclear power plants, if safety concerns can be resolved

The urgency of these proposals, coupled with more immediate evidence of fuel shortages, such as the gasoline shortages of 1973, closing of the Denver school system for part of 1972, and the brownouts and blackouts of recent years, suggest that an energy crisis has certainly arrived. The purpose of this chapter is to explore the dimensions of that crisis and its implications for the supply of electric energy.

This chapter must, of necessity, consider all forms of energy, not just electric energy. First, the question of energy resources itself demands close study, since almost all of our activities are related to energy use in some form. Second, all of the major sources of energy can be converted into electric energy, as well as used in other ways. An increasing competition will almost certainly develop between electric energy and other forms of energy, produced in many cases by the same primary energy sources.

One point must be clearly understood. We do not face a crisis in energy *presence*. We do face a crisis in energy *availability*. Enormous amounts of energy reach us daily from the sun. Almost unlimited energy is locked in the nucleus of the atom. Energy to supply all of our needs for scores or hundreds of years is available in fossil fuels. Other forms of energy surround us in huge quantities. But, as we argued in Chapter 8, until energy is harnessed—until it is made available in useful form—it is of no value to man. Our crisis today is a crisis of availability of energy in a useful form at the right place.

Before we turn to the problem of availability, we look first at the question of energy growth and approaches available toward meeting that growth. We then take up the question of the availability of the various forms of energy. Finally, we consider how we use energy today and what future use we may anticipate in the light of energy availability.

13.1 THE DIMENSIONS OF ENERGY GROWTH

Previously we have worked with energy in units of the BTU or the KWH. As we turn our attention to the study of total amounts of energy needed in a year, or available, say, in coal in the United States, or some other very large quantity of energy, we find these small units inadequate or inappropriate.

Accordingly, we define a new unit of energy, the Q.

$$1 \ Q = 1 \text{ quintillion } (10^{18}) \text{ BTU} = 2.93 \times 10^{14} \text{ KWH}$$

To introduce our use of this new unit we note that the total energy used in the United States in 1970 was about 0.07Q. To give further meaning to this new unit, we indicate in Table 13–1 a number of energy levels expressed in units of Q.

TABLE 13–1.
Typical Energy Levels in Units of Q

Source	Energy in Units of Q
Annual Solar Energy Incident on Earth	5,300
Estimated Uranium Reserves	
With Breeder Reactor	200 to 2,000
With Conventional Reactors	2 to 20
Estimated Fossil Fuel Reserves	15 to 120
World Energy Use (1970)	0.2
U.S. Energy Use (1970)	0.07
U.S. Electric Energy Use (1970)	0.005
One Megaton Nuclear Bomb	0.000004

Next we ask what annual growth and what cumulative demand for energy we may expect in the years ahead. We shall assume that the total energy growth in the United States continues at 3.5 percent for the period of study (electric energy growth is near seven percent). There is, of course, no way to know how realistic this assumption is. It will show us the nature of our energy demand in in the future if such growth continues.

With an assumption of 3.5 percent growth in total energy, the energy in the United States during the year Y is:

$$E_Y = 0.07 \times (1.035)^{Y-1970}, Y \geqslant 1970 \qquad (13-1)$$

The total cumulative energy, E_C, required from the beginning of 1970 to the end of some year Y_F is the sum:

$$E_C = \sum_{Y = 1970}^{Y_F} E_Y$$

$$= \sum_{Y = 1970}^{Y_F} 0.07 \times (1.035)^{Y-1970}$$

$$= 2 (1.035^{Y_F + 1 - 1970} - 1) \qquad (13-2)$$

Equation (13-2) is obtained from the previous line by fairly straightforward methods for adding terms in a geometric series. The formal calculation is left as an exercise for interested students.

E_Y and E_C are plotted in Figure 13-1 to indicate the dimensions of our future needs if we continue at present growth rates. We must remind the reader again that these curves are not predictions. They are projections based on an assumed annual growth rate of 3.5 percent, which is close to the growth rate in recent years. In fact, the actual growth rate may decline because of decreased demand for more technology, or because of environmental pressures, or because of some unforseen major catastrophe. On the other hand, growth rate may also increase for any one of a number of reasons. We are, however, sufficiently certain that growth will continue at least for the immediate future that we must ask whether we will have the energy resources available to meet these needs. Predictions or projections of exponential growth are not new, of course. They are at least as old as Thomas Malthus[1] and as recent as Dennis Meadows.[2]

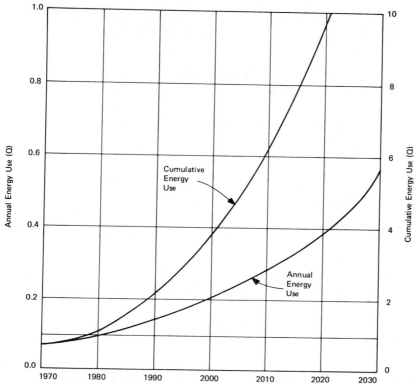

FIGURE 13.1. Projected Future U.S. Energy Needs. This display assumes 3.5% annual growth rate.

The major sources of energy are summarized in Table 13-2.

TABLE 13-2.

Major Sources of Energy

Non-Depletable Energy Sources	Depletable Energy Sources
Water (hydroelectric)	Coal
Geothermal power	Oil
Solar power	Natural gas
Tidal power	Uranium
Fusion (with Deuterium)	
Wood	
Refuse	

Sources are listed as depletable and non–depletable. In some cases, the listing is a bit arbitrary. For example, geothermal energy originates with the heat in the earth's interior, which is certainly finite. However, this heat is so huge that we cannot conceive of depleting it. Similarly, the supply of Deuterium has been estimated as lasting something like 10^9 (one billion) years. But this number is so absurdly large that we list Deuterium as non–depletable. Wood is listed as non–depletable because it can be recovered as a forest in perhaps 10 to 100 years. It is conceivable that wood might be continuously farmed and used for a continuous supply of energy, but this is not commonly done in large quantities. Refuse will be a source of energy as long as man continues to discard burnable materials such as paper, wood, oils, and organic wastes.

We shall start our discussion below with the non–depletable resources. Today they supply much less of our total energy or electrical energy needs than the depletable resources. It seems likely, however, that eventually the non–depletable resources must supply the greater part of our needs. (This conclusion may not necessarily be valid for many hundreds of years if the breeder reactor is developed in a safe form.)

13.2 AVAILABILITY OF NON–DEPLETABLE ENERGY RESOURCES

When we quantify resources that are not depleted, it is not useful to try to talk about the total quantity of energy available, since this is theoretically unlimited. We are, however, interested in the rate at which these resources can be used. Hence we use the dimensions of power in this section, and then return to dimensions of energy in the following sections on depletable resources.

For reference purposes the electric power capacity in the United States in 1970, and the projected capacity for the year 2000 (assuming an annual electric power growth rate of seven percent) are about:

1970	340,000 MW
2000	2,700,000 MW

Of the non-depletable energy sources listed in Table 13-2, only two presently produce an appreciable amount of commercial electric power in the United States:

Hydroelectric   55,000 MW
Geothermal   300 MW

We now briefly review the prospects for each of the non-depletable sources in the future.

1. Hydroelectric. The Federal Power Commission has estimated that there is an additional hydroelectric potential of about 126,000 MW which could conceivably be developed. (3) Much, however, would be far too expensive, or more desirable for other uses, and will never be developed. It is estimated that by the year 2000 the hydroelectric capacity in the United States will have increased by about 70,000 MW to a total of 125,000 MW. This will be a small fraction (about seven percent) of the needs of the year 2000. It should be noted that much of this hydroelectric potential will be in pumped storage. This, of course, implies a depletable resource, since the stored water is pumped "uphill" by power from a thermal power plant which ordinarily uses depletable fuels.

2. Geothermal. The most optimistic estimates suggest that geothermal power may reach 25,000 MW in the foreseeable future. (4) This would be about one percent of our needs in 2000. Actually it is very difficult to predict how big a role geothermal power may play in the future. There are huge heat resources inside the earth. Whether we will find a way to tap substantial quantities of this heat is something only time will tell.

3. Solar. An average of 2,500 MW is incident on each square mile of earth where the skies are usually clear of clouds. Clearly, if an efficient and economical solar conversion scheme can be developed, and if enough square miles of land can be dedicated to the project, any portion of our power needs of 2000 can be met by solar power. It is not clear today whether we can develop an acceptable system or whether we wish to dedicate the necessary land. Solar power appears to have some prospects for major development but it is again too early to be sure of its future.

4. Tidal. There will probably be many proposals to develop one site or another in the years ahead. Capital cost will be a very strong

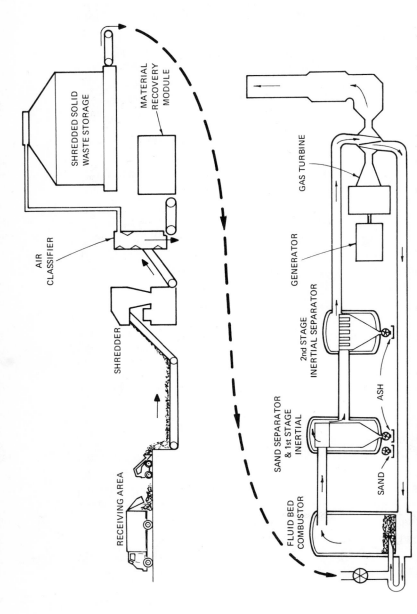

FIGURE 13–2a. Schematic drawing of a system for converting raw garbage into heat and reclaimed materials. The heat from the burning garbage is concentrated in gases which drive a gas turbine-generator powerplant. (*Courtesy Combustion Power Company, Inc.*)

RECEIVING AREA

SHREDDER

AIR CLASSIFIER

SHREDDED SOLID WASTE STORAGE

MATERIAL RECOVERY MODULE

FLUID BED COMBUSTOR

SAND SEPARATOR & 1st STAGE INERTIAL

SAND

ASH

2nd STAGE INERTIAL SEPARATOR

GENERATOR

GAS TURBINE

FIGURE 13–2b. A total refuse–reclamation system. In this photograph the last four stages of power generation in the Combustion Power Company CPU–400 system are shown. (Refer to Figure 13–2a.) At left is the fluid bed combustor. The sand separator and first stage inertial separator is deep in the center; the second stage inertial separator is the smaller container at center right. The gas turbine–generator unit appears at the right. Generating electricity is only a part of recycling wastes. Garbage contains valuable metals such as steel and aluminum which are well worth salvaging. The energy cost of producing usable steel or aluminum from waste material is much less than that of mining and refining the metals "from scratch." Here a separation system using magnetic and air classifying and separating is a valuable component of a reclamation plant. Electric energy generation plus metal recovery could turn garbage-burning into a profit–making enterprise. (*Courtesy Combustion Power Company, Inc.*)

deterrent to development. It is difficult to believe that any substantial amount of our electric energy needs will ever be met by
tidal power.

5. Fusion. Fusion power, along with solar power, offers the major
hope for significant quantities of electric power from non–depletable resources. It seems likely that physical and technological problems will prevent or delay major development at least until around
the year 2000. But there is a real prospect that fusion power may
provide a long range solution to the energy crisis.

6. Wood. Although the burning of wood can provide a great deal
of power for short periods, the time required to replace trees
makes the realization of large amounts of average power unfeasible.

7. Refuse. Recently, a number of new proposals for burning solid
wastes have been made. Some small plants are in operation or
under design. One plant in St. Louis burns 300 tons of municipal
waste per day to generate 12.5 MW.(5) It has been estimated that if
10 percent of the 1970 solid waste had been incinerated, 25 billion
KWH could have been generated; that is about 1.8 percent of the
total 1970 electric energy consumption. (6)

We conclude this section with the observation that some non–
depletable energy resources offer some hope of significant future
exploitation. Today, however, most of our electric energy (about
84 percent) is generated through the consumption of depleting fuel
resources. We turn now to a consideration of these resources.

13.3 THE USE OF ENERGY RESOURCES

To gain perspective for our study of depletable energy resources,
we consider first the amount of energy of each major type used in
the United States in 1972. This is summarized in Table 13-3.(7)

Table 13-4(7) shows where the energy was used by major sectors of the economy.

Detailed data are not yet available on how the major energy
sources were used in 1972. We can get a rather accurate general

picture of energy use, however, by returning to 1969 figures.(8) Table 13–5 shows how each of the three major fuel forms were used in 1969 in percent of total use of that resource.

TABLE 13–3.
Use of Energy Resources in 1972

Energy Source	Amount	Energy (units of Q)	Percent of Energy
Petroleum	5,960 million barrels	0.0328	46
Natural Gas	22,607 billion cu. ft.	0.0233	32
Coal	517 million short tons	0.0124	17
Hydro	280 billion KWH	0.0029	4
Nuclear	57 billion KWH	0.0006	1

TABLE 13–4.
Energy Use by Economic Sectors in 1972

Sector	Percent of Use
Industry	28.8
Electric Energy Genration	25.6
Transportation	25.0
Household and Commercial	20.6

Finally, Table 13–6 shows how energy resources were used in the generation of electric energy in 1972.

With these data as background we turn to the future. What is the probable availability of energy resources for the next few decades?

13.4 THE PHENOMENON OF DEPLETION

The world has only a finite supply of its depletable resources. In time our coal, oil, natural gas, and every other resource we use must be exhausted or become too expensive to extract from the earth. The phenomenon of depletion is studied in an outstanding

TABLE 13-5.

Uses of the Three Major Fossil Fuels (1969)

Petroleum	Automobiles	27.7%
	Other transportation	25.2
	Space heating	20.0
	Process heat	10.9
	Petrochemicals	10.5
	Electric energy generation	5.7
		100.0
Natural Gas	Process heat	40.2%
	Space heat	22.0
	Electric energy generation	17.1
	Water heating cooking	10.1
	Petrochemicals	2.6
	Losses	8.0
		100.0
Coal	Electric energy generation	56.0%
	Process heat	21.7
	Iron and steel production	17.2
	Space heat	3.9
	Petrochemicals	1.2
		100.0

TABLE 13-6.

Resource Use for Electric Energy Generation, 1972

Resource	Percent
Coal	46
Natural gas	23
Hydro	16
Petroleum	11
Nuclear	4
	100

essay by M. King Hubbert recommended to every reader of this text.(9) Hubbert develops in mathematical and historical detail the concept of a resource production cycle.

An idealized example of such a cycle is shown in Figure 13-3. The rate of production of a resource, such as barrels of oil per year, or ounces of silver per year, etc., is plotted against the year of production. If a certain finite amount of a resource exists, it is

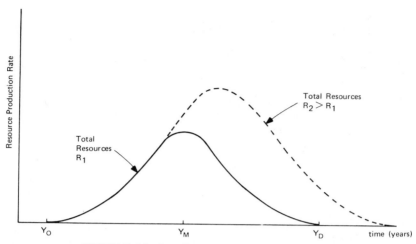

FIGURE 13–3.   Resource Production Cycle.

hypothesized that its production rate will follow, approximately, the curve shown in dark in Figure 13-3. If a greater amount of the resource exists, then the dashed line is more accurate. In practice, production rate curves are not smooth as shown here but are somewhat jagged. Nonetheless, the examples shown by Hubbert fit the pattern in general shape quite well. (9)

Production begins in year Y_O. In some cases, such as that of coal, Y_O is lost in antiquity. In other cases, such as the California gold rush of 1849, or the production of Uranium, Y_O is known accurately.

The year of maximum production is Y_M. In some cases, such as that of natural gas, we may have already reached Y_M. For oil it is probably some years into the future. For coal or Uranium it may be generations from today.

Y_D is the year of depletion. It is not a well-defined date, since small pockets of material may continue to be found almost indefinitely. In the foothills of the Sierra Nevada, campers, weekend miners, or small mining companies still find an occasional nugget or a few flakes of gold even though the peak production of gold occurred over 100 years ago.

The logic of the resource production cycle is basically quite simple. In the early years of the cycle, production grows almost exponentially as demand increases continually. Eventually a point is reached where the cost of extraction begins to go up. Growth rate slows. Eventually, with ever increasing costs of extraction as the easy "near-surface" resources are used up, the production rate reaches a maximum, and then decreases toward zero. This does not mean that all of the resource has been used, but rather that it is no longer economically feasible to extract the remaining resource.

The symmetry of the ideal curve of Figure 13-3 is upset by the discovery of new reserves. The Alaska oil find is an excellent example of an unexpected new reserve. It is common to classify reserves in terms of estimates of the ease of extraction and the state of knowledge about the existence of the reserves. There are a number of degrees of refinement of these estimates.(8) We shall consider briefly a fairly simple model and then further reduce the complexity of the estimation problem for our study of specific fuels. Figure 13-4 suggests one way of considering resources.

The reader should realize that the lines suggested are very flexible, if not amorphous. What is possible or undiscovered today may

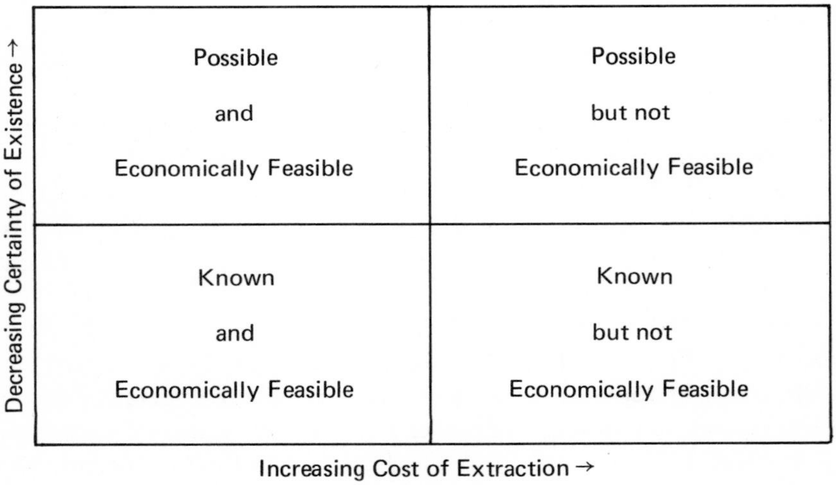

Possible and Economically Feasible	Possible but not Economically Feasible
Known and Economically Feasible	Known but not Economically Feasible

Decreasing Certainty of Existence →

Increasing Cost of Extraction →

FIGURE 13—4. Estimates of Resources.

be known tomorrow. What is financially infeasible at one time may become feasible in a new economic environment.

The relatively simple model of Figure 13-4 is valuable conceptually but is still a little bit complex for the estimates we shall be making in the sections to follow. Accordingly, we reduce the model to a very simple two-part form. The lower left block, called *Known and Economically Feasible* (resources) will be called *Proved Reserves*. The remaining three blocks will be referred to as *Possible Resources*.

Changes in knowledge about resource availability and feasibility of recovery tend to make the right side of the curve in Figure 13-2 irregular and unpredictable. The ideal model is a first order representation of how the production rate tends to vary with time. To get a more detailed picture of what the future may hold we must look at specific fuels and try to estimate what reserves may exist and the costs of these reserves.

In the sections to follow we consider petroleum, natural gas, coal, and Uranium. This is the order of their present use level as general energy resources, though not as fuels for electric energy generation. (See Tables 13-3 and 13-6.)

13.5 PETROLEUM

As Table 13-3 indicates, almost six billion barrels of petroleum were consumed in the United States in 1972. This implies an average of about 16 million barrels per day. Most of this is crude oil which is refined into a large number of petroleum products. Gasoline used in transportation, along with residual fuel oils used in oil-burning electric generating plants, are of particular importance.

The sources of the 16 million barrels used each day in 1972 in the United States are given in Table 13-7. (7)

The demand for petroleum is growing at about four percent per year.(4) This implies a United States petroleum demand over the next decade or so approximating that shown in Table 13-8.

TABLE 13-7.

Sources of Petroleum Used in the United States

Source	Million Barrels per Day
Domestic (U.S.) production	11.6
Western Hemisphere imports, primarily Canada and Venezuela	3.2*
Eastern Hemisphere imports	1.2

TABLE 13-8.

Estimated United States Petroleum Demand

Year	Million Barrels per Day
1972	16
1975	18
1980	22
1985	27

The sources of our petroleum needs for 1972 were shown earlier. It is reasonable to ask what our sources may be in the future. To get some perspective on this, we turn to the question of where the world's petroleum reserves are to be found, as shown in Table 13-9.(4, 10) Of the six billion barrels of oil consumed in the United States in 1972, about 4.4 billion barrels came from domestic reserves and 1.6 billion barrels from imports. It is apparent that our present known reserves of 29 billion barrels are not large in the context of a six billion barrel–per–year demand. We have essentially three alternatives: (1) decrease demand; (2) expand our known reserves, or (3) import more petroleum.

The first alternative was not widely or seriously considered until late 1973 with the advent of the Arab states' embargo. The petroleum energy policy or strategy of the United States will almost certainly center for the time being on the second and third alternatives.(11)

---

*Much of this actually consists of petroleum products refined elsewhere (*e.g.*, in Holland) from Middle East crude oil. This explains why the Arab embargo of late 1973 suddenly caused a deficiency of about 17% of U.S. needs.

TABLE 13-9

World Petroleum Resources

Geographical Location	Proved Reserves (billion barrels)	Possible Resources* (billion barrels)
Middle East	345	800
Communist World	100	500
Africa	75	250
Continental U.S.	29	420–2,700
Latin America	26	225
Alaska	15	30–35
Europe	14	20
Far East	14	200
Canada	11	95

*Estimates based on References 4, 8, and 9.

On April 18, 1973, President Nixon sent to the Congress an energy message dealing with most major aspects of the energy problem. With respect to petroleum, the policy encouraged imports in the short term (the mid–1970's) but reliance on domestic resources in the long term. The latter will require development of new reserves. In fact, there is little else the United States can do in the short run except increase imports, short of restricting or rationing fuels. This country does not presently have the physical facilities required for major production expansion.

To implement the petroleum policy, the President issued an Executive Proclamation which removed existing tariffs on crude oil and petroleum products. This has the effect of encouraging imports to meet demand. At the same time a license fee quota system was introduced. This permits holders of import licenses to import petroleum without fee up to the level of their 1973 quota allocations. Beyond that level an import fee would be charged. This fee would have the effect of encouraging domestic production.

So far in this section we have seen the demand for oil, the world's sources of oil, and a petroleum policy for the United

States. We consider first the prospects of meeting the import objectives, and second the prospects of expanding domestic production.

Prospects for meeting import objectives are of course closely related to world petroleum resources (Table 13-9). It is necessary that we consider the proved reserves column, since the possible resources column is highly speculative. If we are to meet the demand projected in Table 13-8, imports will have to come from sources other than Canada or Latin America. The only presently credible source is the Middle East. Imports from the Middle East in 1985 could range from 10 to 15 million barrels per day at a cost close to $20 billion per year. Imports of this magnitude have major economic and political implications. The cost would represent a huge increase in our balance-of-payments deficits.

Political problems may be even more serious. The relations of the United States with the Arab nations are not particularly stable. The Middle East has oil and wants the income it produces. It also has a certain degree of independence in selling that oil. Europe and Japan are major and growing customers for Middle East oil. The United States, as a protector of Israel, may not be an entirely welcome newcomer as a major customer. In addition, the competition for limited reserves may lead to new frictions among European nations, Japan, and the United States.

Considerations such as these suggest that we turn our attention to the possible expansion of domestic supplies. Perhaps no petroleum resource has gained more attention than the north-slope Alaskan oil discovery of the last decade. Reserves are listed as 10 to 15 billion barrels with a possible 35 billion or more additional barrels. It has been proposed that an 800 mile pipeline may be built from the main reserves at Prudhoe Bay in Northern Alaska. This pipe would deliver about two million barrels per day in 1980, or about 10 percent of demand. Construction of the pipeline, which had been stopped temporarily by environmentalists who fear damage to the Alaskan tundra, was authorized by Congress in July, 1973. An alternative, and much longer, line would cross Canada and enter the midwest. Proponents of the shorter Alaska line argue that it can be built in less time for much less money. However, it also should be noted that if the oil is piped to a sea-

FIGURE 13–5. A supertanker. Supertankers like *Esso Scotia* are carrying an increasing part of the crude oil that moves between countries. At 250,000 tons, *Scotia* is 100 feet longer and 50 feet wider than the largest passenger liners built; crew members ride bicycles on her deck. Very large oil tankers offer major savings in economics of scale of transportation; such savings are imperative when we are importing quantities of petroleum. *Esso Scotia* carries almost two million barrels of oil. This is still only enough to meet U.S. petroleum demand for about three hours. Few ports in the United States can handle such large ships. Major new imports would require construction of many additional off–loading facilities. A tanker of conventional size, *Esso York*, is shown alongside *Scotia* taking on a load of oil for transshipment to a shore facility. (*Courtesy Exxon Corporation.*)

port such as Valdez, it can be sold to the highest bidder, which might well be Japan.(12)

Another major source of future petroleum supplies is the Outer Continental Shelf (OCS) off the coast of the United States. In his message President Nixon urged extended leasing and exploration in this area.

Whether these sources plus a number of other potential sources on shore in the coterminus 48 states can be developed to supply the needs of the future is yet to be seen. But that is now the policy of the United States.

Before we leave the subject of oil we turn briefly to costs of oil over a brief time span. Today costs are rising quite rapidly for a number of reasons, including greater competition, increasing production costs, and inflation. The following data (based on References 13 and 14) give an approximate idea of the cost of crude oil per barrel.

1963	East coast import	$1.90/barrel
1973	East coast import*	4.40
	Average domestic production	4.00
1985	Average domestic production	7.00

It should be understood that today's prices change almost weekly, and the estimate for 1985 is highly speculative.

In Section 13.9 we consider some additional sources of petroleum: oil shale and tar sands.

13.6 NATURAL GAS

The problem of meeting natural gas needs is significantly different from that of meeting petroleum demand. Natural gas reserves are much smaller than petroleum reserves, and the prospects for intercontinental imports much more difficult.

Table 13–3 indicates that almost 23 trillion cubic feet (TCF) of natural gas were consumed in the United States in 1972. The total proved reserves at the end of 1972 are estimated at about 280 TCF. This suggests that we have about enough natural gas left for the next 12 years. However, estimates of potential recoverable resources of gas range from 800 to 2,100 TCF. This is equivalent to a resource lifetime, at current use, of from 40 to 100 years. (4, 8)

Proved reserves (excluding the Alaskan reserves—about 30 TCF) have been going down since about 1968. Exploration efforts have also declined. Why should efforts to find new resources decline in the face of the rapidly approaching reserve depletion? The answer is closely related to the cost of natural gas in comparison with other fuels. In order to make such a comparison we express costs per million BTU's for the major fuels in Table 13–10.(13)

---

*By the end of 1973 Arab producers had raised their price to $11.53.

TABLE 13-10.

Approximate Fuel Costs (1973)

Fuel	Cost per million BTU's
Natural gas (wellhead)	$.25
Coal	.35
Crude oil	.60
Heating oil	.80

The cost of natural gas at the wellhead (place of extraction) has been regulated by the Federal Power Commission at $.25. This has made natural gas the least expensive fossil fuel. The result is that it has been used very extensively, and, some contend, extravagantly. The gas industry has argued that the price is artificially low and does not justify further exploration. Others have argued that the regulated price was adequate compensation for the producer, with sufficient income to permit exploration.

In his energy message, President Nixon proposed legislation to remove controls on prices of new supplies of natural gas. (11) If enacted, this legislation will permit gas to "float" in the energy market. Its price will probably double (approximately). The result should be a more balanced use of fuels, along with increased exploration for reserves. This should lead to the availability of domestic natural gas for many more years.

Natural gas is imported into the United States from Canada through pipelines. It is not feasible, however, to transport gas in its gaseous state by ship from major reserves in South America, Africa, Russia, and the Middle East. If shippers liquefy gas by cooling and compressing it, they can ship it overseas in specially-constructed vessels, something like floating thermos bottles. Liquid natural gas (LNG) could be imported for about $1.25 to $1.50 per million BTU. This price is presently somewhat above the market price for fuels. LNG may someday supply significant amounts of energy. If, however, domestic reserves do respond to price incentives, imported LNG may not be competitive on the energy market.

FIGURE 13–6. Fossil-fuel plant at Moss Landing, California. At 2,100 MW, the Moss Landing Power Plant, on the Pacific Ocean 100 miles south of San Francisco, is the largest in the United States. It was completed in 1968 at a capital cost of $214 million; its cost per kilowatt is a very attractive $100. Moss Landing provides an excellent example of a two–fuel fossil plant. It has had to switch from natural gas to residual oil because of the growing shortage of natural gas. (See Chapter 13.) In 1970, 98% of Moss Landing's energy came from natural gas. Only 2% came from oil; the use of oil occurred during cold spells in the winter when demand for gas for heating purposes normally increases. By 1975, Moss Landing will have converted almost entirely to residual oil. With the conversion, unfortunately, will come increased air pollution problems. (*Courtesy of Pacific Gas and Electric Co.*)

13.7 COAL

Coal is by far the most abundant fossil fuel in the United States. Until the breeder reactor is successfully developed, coal will represent over three fourths of our fuel reserves. It is also the dirtiest of fuels, so it requires extensive and costly measures to make its use environmentally acceptable. The problem with coal is not to find it but to use it cleanly.

In 1972 the United States consumed 517 million short tons of coal. Proved reserves are estimated at about 400 billion tons.(4) Possible resources are as much as 3,200 billion tons, with perhaps half of that recoverable.(15) This suggests resources which might last for many hundreds or perhaps thousands of years. But before we can talk about this magnitude of use we must consider the problems associated with using coal.

Coal is found in the eastern United States along the Appalachians, in some Midwestern states (Illinois, Ohio, Indiana and others), and in or near the Rocky Mountain states (Montana, Wyoming, Colorado, New Mexico). Many other states have lesser reserves, though often enough to justify some power development. An example is Washington State with its new Centralia plant. (Chapter 5)

Roughly half of the coal in this country is on each side of the Mississippi. Eastern and Midwestern coal in states such as West Virginia, Pennsylvania, Kentucky, Illinois and Ohio has been mined extensively for many years because of its proximity to large concentrations of industry and population. Unfortunately, coal in the eastern part of the United States tends to have rather high sulfur content, while western coal has much less sulfur on the average. With increasingly strict air pollution control laws, interest in coal is shifting from east to west.

States such as North Dakota, with 50 billion tons of low sulfur strippable coal, Montana with 15 billion tons, and Wyoming with 10 billion tons all appear to be ideal sources of fairly clean energy, with lots of open space and few neighbors to object to a local power plant. (16) But the lessons of Four Corners are not easily forgotten.(17) The air pollution which coal–fueled power plants brought to the Southwest could become a fact of life in the northern Rocky Mountain states. Such moves will probably be opposed by local residents.(18) Stripping and burning the coal of Montana for the energy needs of Chicago will no doubt meet increasing opposition in Montana if not in Illinois.

Air pollution is not the only concern in coal–rich states. Most of the coal under consideration for development would be strip mined. An overburden of dirt up to perhaps 100 feet in thickness is removed. The coal seam is then removed by huge stripping machines. Strip mining has two major advantages over deep mining. It is easier and less expensive, and it is far less dangerous to miners. But it has very important disadvantages. If the exhausted coal seam is left exposed, the result is an ugly scar on the land. Also, rain may drain mine wastes into local watersheds and rivers, causing severe pollution in many cases. Much of this problem is eliminated by returning the overburden of dirt which was originally stripped off and replanting the land. However, it is necessary that the topsoil of the replaced overburden be good growing ground. (19)

FIGURE 13–7. After the strip–mining. When the overburden is removed and coal is stripped from the earth, the land should be restored by re–seeding and planting of new trees. This helps prevent erosion and leaching of mine wastes into streams, and it can restore some of the natural appearance of the land. Here a crew following a strip–mining "bench" plants loblolly pine seedlings. (*Courtesy of the Tennessee Valley Authority.*)

Coal need not be used directly, as it is found, for energy production. One alternative which should produce a much cleaner and easily transported fuel is coal gassification. This process converts coal into a gas which has many of the advantages of natural gas, and could replace it as supplies of natural gas diminish. El Paso Natural Gas and Texas Eastern Transmission Corporation are planning plants which would each produce about 250 million cubic feet per day. This represents a little less than one percent of the natural gas consumption rate in 1972. This gas should cost about $1.20 per million BTU's, which could be a little high to be competitive. Other less expensive systems are under study. Gas from coal could produce as much as four percent of our energy needs by 1985, and 25 percent by 2000, if adequate technologies can be developed.(13)

The cost of coal has been rising fairly rapidly. In 1929 TVA paid $2.69 per ton. In 1971 it was $5.90 and long term contracts call for costs close to $8 per ton.(19) Increased concern for safety, environmental protection of the land, and increasing inflation will no doubt cause coal prices to continue to rise, probably rather significantly for at least the immediate future.

13.8 URANIUM

Reserves of Uranium in terms of remaining years of supply are very difficult to estimate for a number of reasons. First, the nuclear industry is small but growing rapidly. We shall adopt here the common assumption that nuclear power will provide about 50 percent of our electric energy needs by the year 2000. Second, Uranium can be extracted at a wide variety of price ranges. The maximum feasible extraction cost is very difficult to estimate today. Third, successful development of the breeder reactor would make fuel costs almost meaningless, and it should also extend fuel lifetime into the thousands of years.

Uranium ore is mined and processed into U_3O_8 which is the form for which costs are generally stated. The cost of U_3O_8 is presently about $7 to $8 per pound. It is generally estimated that the supply of "$8 Uranium" will last about 20 to 30 years at projected growth rates. It is sometimes suggested that we have only about 30 years worth of Uranium left. But this is deceiving. As Table 13-11 suggests, the amount of U_3O_8 available is highly dependent on the cost of extraction. We use here the terminology of the original document.(20)

The column on the right in Table 13-11, giving the incremental cost of energy due to increased U_3O_8 cost, assumes an average energy cost increment of 0.06 mills per dollar per pound of U_3O_8. (See also Table 5-1 and Problem 5-15.) The message of Table 13-11 is that while $8 Uranium may be in fairly short supply, much more fuel is available at prices which are probably not prohibitively high.

TABLE 13-11

Estimated U.S. Uranium Reserves—January 1, 1970

U_3O_8 Price Per Pound	Reasonably Assured (Cumulative) (tons)	Estimated Additional (Cumulative) (tons)	Total (Cumulative) (tons)	Increase in Energy Cost Due to Higher Fuel Cost (mills/KWH)
$ 8.00	204,000	390,000	594,000	0.0
10.00	340,000	600,000	940,000	0.1
16.00	500,000	950,000	1,450,000	0.4
30.00	640,000	1,600,000	2,240,000	1.3
50.00	6,000,000	4,000,000	10,000,000	2.5
100.00	12,000,000	13,000,000	25,000,000	5.5

For example, the total supply of Uranium increased by about four times for an energy cost increase of about 1 mill/KWH. It seems likely that other factors in the complex and changing energy market will override this relatively small change.

If, of course, the breeder reactor is developed, the situation changes dramatically, since the fuel–use factor goes up by about 100 times with the utilization of U^{238}. Recall from Chapter 6 that over 99 percent of natural Uranium is fertile U^{238}, which can fission in the breeder, while only 0.7 percent of natural Uranium is fissile U^{235}.

13.9 OIL SHALE AND TAR SANDS

We saw in Section 13.5 that proved oil reserves are not particularly large in terms of our present use. Possible resources are much more significant but remain to be proved. Two additional sources of petroleum exist with some potential for meeting our energy needs. They are *oil shale* and *tar sands*. Reserves are large, but we do not as yet have the technology for extensive development.

Oil shale is a fine–grained sedimentary rock that contains enough organic matter to yield 10 gallons of oil or more per ton.(21) Reserves are estimated at about two trillion barrels of oil, about equal to the total possible resources of petroleum in the United States. A relatively small fraction of this is probably recoverable. One estimate of shale production cost is $3.90 per barrel.(13) The extensive mining required and the huge wastes produced will pose major environmental challenges to development.

Tar sands are asphalt–bearing rocks which are a source of synthetic crude oil. The Athabasca deposit in Alberta, Canada is the world's largest. It may provide one of the best sources of petroleum for the United States. Reserves are estimated at about 300 billion barrels of oil. Production cost should be about $4 per barrel.

13.10 SOME COMMENTS IN CONCLUSION

With the array of statistics of knowns and unknowns, we are left still with the question of which energy sources will meet our needs in the decades ahead. Recognizing the dangers inherent in predicting the future, we nonetheless suggest some fairly reasonable prospects for the next few decades.

1) For the next seven to ten years, fossil fuels, essentially as we use them today, must provide most of our energy needs. Petroleum imports, as well as domestic exploration for oil and gas, will increase.

2) 1980–1990. A large number of conventional nuclear re-

actors will come on line and nuclear power will begin to provide a significant amount of energy, unless major flaws in nuclear safety develop. Fossil fuel use will become more internalized, with fewer imports. Geothermal power may begin to contribute to the nation's energy needs.

3) 1990–2000. Conventional nuclear power will begin to dominate the electric power field. The breeder reactor will begin to contribute significantly to energy needs. Domestic fossil fuels will continue to be used extensively. Geothermal and solar power may begin to be significant if development is successful in the period 1973–1985. Nuclear fusion may be shown feasible and be in the early stages of development, but will not contribute any significant energy in this decade.

In this chapter we have talked about the energy resources required to meet the demands of society. We shall delay until the last chapter the question of how we might conserve energy and resources.

REFERENCES FOR CHAPTER 13

1. T.R. Malthus, *First Essay on Population*, 1798, *Reprints of Economic Classics*, A.M. Kelley, New York, 1965.

2. D. Meadows *et. al., The Limits to Growth*, Universe Books, New York, 1972.

3. "Hydroelectric Power Resources of the United States—Developed and Undeveloped," Federal Power Commission, January 1, 1968.

4. *U.S. Energy—A Summary Review*, Department of the Interior, Washington, D.C., January, 1972.

5. *Energy Recovery from Waste*, U.S. Environmental Protection Agency, 1972.

6. E. Hirst and T. Healy, "Electric Energy Requirements for Environmental Protection," *Public Utilities Fortnightly*, Vol. 91, No. 10, May 10, 1973, pp. 52-60.

7. "The President's Energy Message, Summary Outline—Fact Sheet," Office of the President, Washington, D.C., April 18, 1973.

8. *Reference Energy Systems and Resource Data for Use in the Assessment of Energy Technologies*, for Office of Science and Technology, Associated Universities, Inc., Upton, N.Y., April, 1972.

9. M. King Hubbert, "Energy Resources," in *Resources and Man*, National Academy of Sciences—National Research Council, W.H. Freeman and Co., San Francisco, Calif., 1969.

10. R.A. Rice, *The Transportation of North Slope Oil and Long-Range Alaska Transport Needs*, Carnegie Press, Pittsburgh, June, 1972.

11. R.M. Nixon, Energy Message to the Congress of the United States, the White House, Washington, D.C., April 18, 1973.

12. A.M. Louis, "The Escalating War for Alaskan Oil," *Fortune*, July, 1972.

13. "Enough Energy if Resources are Allocated Right," *Business Week*, April 21, 1973, pp. 50–60.

14. "Growing Drive to Tap Vast U.S. Fuel Reserves," *U.S. News and World Report*, April 30, 1973, pp. 69–76.

15. P. Averitt, *Coal Reserves of the United States, Geological Survey Bulletin* 1275, U.S. Government Printing Office, 1969.

16. *The Economy, Energy, and the Environment*, Joint Economic Committee, Congress of the United States, U.S. Government Printing Office, Sept. 1, 1970.

17. A. Wolff, "Showdown at Four Corners," *Saturday Review of the Society*, June 3, 1972, pp. 29–41.

18. S. Jacobsen, "The Great Montana Coal Rush," *Science and Public Affairs*, April, 1973, pp. 37–42.

19. W. Greenberg, "Chewing it up at 200 Tons a Bite: Strip Mining," *Technology Review*, February, 1973, pp. 46–55.

20. *Potential Nuclear Power Growth Patterns—WASH 1098*, Atomic Energy Commission, U.S. Government Printing Office, Washington, D.C., December, 1970.

21. "Oil Shale—A Potential Source of Energy," U.S. Dept. of the Interior, U.S. Government Printing Office, Washington, D.C., n.d.

GENERAL READING FOR CHAPTER 13

1. *Resources and Man*, National Academy of Sciences—National Research Council, W.H. Freeman and Co., San Francisco, 1969.

   This is an outstanding book on resources in general and on their relation to man's life. The chapter written by Hubbert (reference 9) is one of the classic references on energy resources.

2. E. Ayers and C.A. Scarlott, *Energy Sources—The Wealth of The World*, McGraw–Hill Book Co., New York, 1952.

   Although this book is over 20 years old, much of the material is timeless. The first chapters on the history of energy are fascinating. In later chapters the numbers may be a bit off but the ideas are as relevant as they were 20 years ago.

3. *Reference Energy Systems and Resource Data for Use in the Assessment of Energy Technologies*, for Office of Science and Technology Associated Universities, Inc., Upton, N.Y., April, 1972.

   This is a very complete and up to date summary of energy sources and energy use. It is not meant for casual reading, but it is a wealth of important data.

PROBLEMS FOR CHAPTER 13

General Problems

13.1. Express the electric energy used in the year 1970 in units of Q (Hint: see Chapter 1.) (0.005)

13.2. Discuss in what sense the following energy sources may or may not be considered non–depletable: geothermal; tidal; wood.

13.3. If 10,000 MW of geothermal power can be developed by 1982, what will be the approximate percentage of capacity in that year represented by this source? (1.4%)

13.4. Use Tables 13-3 and 13-5 to determine approximately how many million barrels of petroleum were used for automobile propulsion in 1972.

13.5. Make a sketch of the general shape of a resource production cycle curve if a major new easily extracted reserve is located some years after Y_M.

13.6. If the domestic production of oil in the United States (excluding Alaska) continued at the level shown in Table 13-7, how long would our proved reserves last? (6.7 years)

13.7. Repeat Problem 13.6 if Alaskan reserves are included. (9.4 years)

13.8. Explain why the statement "We have only 12 years worth of natural gas left" is misleading.

13.9. Estimate the volume in cubic miles of coal used in the United States in 1972.

13.10. Why is it almost certain that most of our energy needs must be supplied by fossil fuels over the next decade?

Advanced Mathematical Problems

13.11. Derive Equation (13-2) from Equation (13-1).

13.12. Use Equation (13-2) to estimate the cumulative energy use between 1970 and 2000.

13.13. Assume that an ideal resource production cycle can be accurately modeled by a Gaussian probability law of mean Y_M and standard deviation Y_S. Write the equation for the production rate. Draw the curve of the cycle. If Y_S is 40 years, how many years after Y_M is the resource 90% depleted?

13.14. For the assumptions made in Problem 13.13, when is the production rate changing most rapidly?

13.15. Write an equation for the increase in nuclear energy cost versus U_3O_8 cost under the assumption made for Table 13-11. Find the cost of U_3O_8 which corresponds to an energy increase of 2 mills/KWH. ($41)

Advanced Study Problems

13.16. Discuss the sources of oil presently used in the United States in terms of country of origin, costs, political considerations, amounts, environmental factors, etc.

13.17. Make a map of the United States indicating where coal deposits are located, what their heating value and sulfur content is, and what the implications are for the electric energy industry.

13.18. On the basis of what you have seen in this chapter and other information available to you, make your own estimate of prospects for energy use in the next three decades, analogous to the estimate made in Section 13.10.

13.19. Compare the advantages and disadvantages of the Alaskan pipeline and the trans–Canadian pipeline as means of bringing Alaskan north slope oil and gas to the United States.

14

ENERGY AND THE AUTOMOBILE

Things are in the saddle,
And ride mankind.

R. W. Emerson
Ode to W. H. Channing

The first chapter of this book was concerned with man's use of electric energy. In the intervening chapters we have dealt with the problem of meeting the demand for electric energy. Chapter 13 was concerned with all energy sources. In this chapter we turn again to the *use* of energy, this time to a specific use, the automobile.

The automobile is an extremely interesting example of energy use at this point for two reasons. First, it uses a very large part of our general energy resources, and it is therefore a subject to consider for energy conservation. Second, it is an excellent example of a device which could be converted from its present energy use to electric energy, if that conversion were deemed desirable.

With what we have learned in Chapter 13, we are prepared to discuss the present energy use of automobiles. With what we have seen in all of the book so far, we are in a position to consider the advantages and disadvantages of the electric automobile.

14.1 THE GASOLINE AUTOMOBILE

From Tables 13-3 and 13-5 we can determine that about 13 percent of all the energy consumed in the United States in 1972 was used in powering automobiles. One recent study indicates that an additional eight percent of our total energy budget was spent on the automobile in indirect ways, including construction, sales, repair, building and maintaining highways, etc.(1) Thus over one fifth of our total energy budget goes into the owning and operating of the private automobile. Most of this energy comes from petroleum.

In this book we limit our discussion to energy use for propulsion. The energy consumption or gasoline mileage of an automobile is a complex function of a number of factors. These include weight, speed, engine size, operating altitude, condition of the car, ability of the driver, power of the engine, etc. A number of recent studies have pinned down the gasoline consumption effects of two of these factors, namely weight and engine size.(2, 3, 4)

Figure 14-1 indicates the relation of energy use to vehicle weight.(2, 3) If we take an average between the two curves, we obtain an equation for energy use in terms of BTU's per pound (of vehicle) per vehicle mile.

$$C_E = 2.5 \text{ BTU/lb./V–m} \tag{14-1}$$

Example 14-1

If a 4,000 pound automobile (just a little over the average weight) is driven 10,000 miles in a year (about average), how much energy does it consume?

$$E = C_E \times W \times M$$
$$= 2.5 \times 4,000 \times 10,000$$
$$= 100,000,000 \text{ BTU's}$$

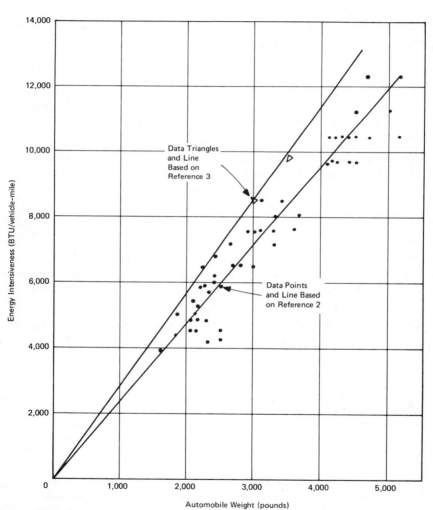

FIGURE 14–1. Automobile Fuel Consumption as a Function of Vehicle Weight (1971 and 1972 cars).

If we assume an electric power plant efficiency of 34 percent (heat rate = 10,000 BTU/KWH), the above energy is equivalent to 10,000 KWH. This is 30 percent more than the average electric energy consumed in the United States per person per year. (See Chapter 1.)

Most people are used to thinking in terms of miles per gallon (mpg) for a vehicle. If we convert Equation 14-1 to this form we obtain an expression for mileage (M):

$$M = 54,400/W \quad \text{(mpg)} \tag{14-2}$$

where W is the weight of the vehicle. This is an average figure and it will be reduced by such factors as air conditioning, excessive speed, stop–and–go traffic conditions, and many more.

But weight alone does not determine energy use. One study relates mileage to cubic inches of displacement of the engine.(4) The data suggest that mileage can be fairly well modeled (or approximated) by the following equation:

$$M = 200/\sqrt{D} \quad \text{(mpg)} \tag{14-3}$$

where D is engine displacement in cubic inches.

The conclusion from this latter study is that it is desirable to cut engine size as well as vehicle weight if energy conservation is a primary objective.

The relation of energy consumption to weight is not unique to gasoline automobiles. We shall see the same effect in the electric car.

14.2 THE ELECTRIC CAR

We face a number of problems associated with the use of the gasoline automobile.

1. It consumes huge amounts of energy from petroleum.
2. It is relatively inefficient compared to some modes of transportation, such as buses, trains, motorcycles, etc.
3. It is a major contributor to air pollution.

We turn our attention now to the electric automobile, asking whether such a vehicle might solve some of these problems.

In the complete electric–car energy system or cycle, energy is found in a number of forms, some of which have been seen in earlier chapters. We start by defining the energy forms of interest to us. The energy forms and interform efficiencies associated with electric cars are defined below and summarized in Figure 14–2. Included in Figure 14–2 in parentheses are approximate efficiencies assumed in this analysis.

KWH_g is the fuel energy in its natural location in the *ground* in the case of the fossil–fuels or Uranium, in a reservoir in the case of hydroelectric power. This fuel is usually called *primary fuel.*

e_p is the efficiency with which the fuel can be removed from its natural state and *processed* appropriately for use in an electric power plant. The range of e_p is quite wide, since this efficiency is highly dependent on the fuel being processed and the effort required by the processor for the particular fuel location. A range of perhaps 60 to 95 percent is not unreasonable. We assume a figure of 90 percent for calculations here.

KWH_t is the *thermal* fuel energy available at the input to the power plant. This fuel is usually called a *secondary fuel.*

e_t is the *thermal* efficiency of the power plant. In modern steam plants it can vary from about 32 percent for light–water nuclear reactors to 40 percent or more for fossil–fueled plants or gas–cooled nuclear reactors. The national average efficiency in 1970 was about 32 percent. We assume a figure of 32 percent.

KWH_e is the *electric* energy available at the output of the power plant. It is sometimes called the bus bar energy. It is the energy which must be produced by any means available to meet consumer demand. Electric energy production is quoted by the industry in units of KWH_e.

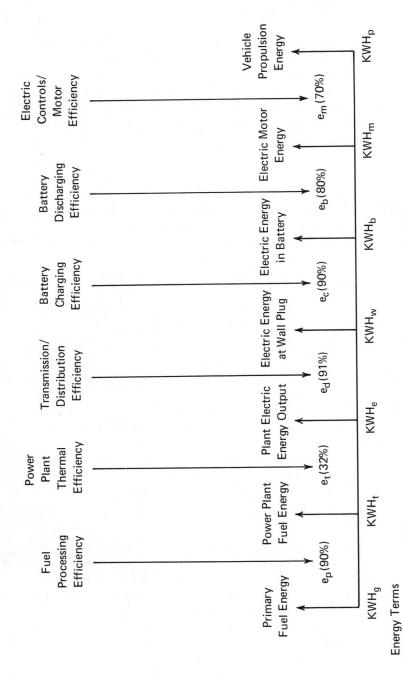

FIGURE 14–2. Energy Forms and Efficiencies Associated with Electric Motor Vehicles.

308

e_d is the efficiency with which electric energy is *delivered* from the power plant to the consumer. It accounts for losses in transmission and distribution. Very long lines or complex distribution systems will reduce e_d. In 1970 the national average was about 91 percent. We assume a figure of 91 percent.

KWH_w is the electric energy available at the consumer's *wall* plug. It is the energy upon which his electricity bill is based.

e_c is the efficiency with which the battery is *charged*. It depends on the nature of the battery, the charger, and time rate of charge. The range of this efficiency is about 85–95 percent. We assume 90 percent here. 85–95 percent. We assume 90 percent here.

KWH_b is the chemical energy stored in the *battery* of the electric motor vehicle. It must be large enough to provide acceptable power and range to the vehicle.

e_b is the discharge efficiency of the *battery*. It decreases with the time rate of discharge. The greater the power demand from the battery, the lower the discharge efficiency. The range on e_b may be about 70–85 percent. We assume 80 percent.

KWH_m is the energy applied to the electric controls and *motor* of the vehicle.

e_m is the efficiency with which the controls and *motor* transform electric energy into mechanical propulsion energy. The probable range is about 60 to 75 percent. We assume 70 percent.

KWH_p is the mechanical *propulsion* energy which moves the vehicle. This energy is eventually dissipated as heat in the wheel friction losses, acceleration and associated friction braking, and in the wind friction of the moving vehicle.

If we multiply all of the efficiencies assumed above we obtain a net efficiency of about 13 percent for the electric automobile. If we consider the probable high and low limits of the above efficiencies we obtain the following range.

Lowest Net Efficiency	*Assumed Net Efficiency*	*Highest Net Efficiency*
6%	13%	21%

14.3 ENERGY REQUIREMENTS OF AN ELECTRIC CAR

Assumptions of the electric energy required by an electric automobile, found in the literature, tend to be fairly consistent in magnitude. However, the use of terms tends to vary. Some authors speak in terms of power plant energy (KWH_e); others work with wall plug (KWH_w) or propulsion (KWH_p) energy. One of the objectives of this section and the last is to refer the estimates made in the literature to a common base. We have chosen as our base the power plant electric energy (KWH_e). This allows us to consider the effect of the growth of use of electric vehicles on the growth of the power industry, since the latter growth is usually expressed in units of KWH_e.

To change from one form of energy to another we divide by the product of all of the efficiencies between the two forms when moving from left to right, and multiply by this product when moving from right to left. (Refer to Figure 14–2.) For example:

$$KWH_t = KWH_b / (e_t e_d e_c)$$

$$KWH_p = KWH_e \times (e_d e_c e_b e_m)$$

Table 14–1 gives the electric energy required to power a vehicle, as determined by actual tests, or as estimated by a number of researchers. Some of the estimates are interdependent. That is, one researcher has relied partially on the results or estimates of others. The degree of interdependence is not clear from the literature. There is, however, sufficient agreement to give one confidence that the estimates are fairly reasonable.

14.4 SCENARIOS OF FUTURE ELECTRIC ENERGY GROWTH

The term *scenario* has become almost a cliché in recent years. A short discussion of whether the term is unnecessary jargon or whether it has an important and significant linguistic role may serve to clarify one of the objectives of this chapter: the consideration of possible futures.

Man has always been interested in "the future." The recent kindling of environmental concerns has accentuated that interest. We are constantly faced with "predictions," "forecasts," "projections," and "estimates." All are based on assumptions, explicit or implicit. The purpose of the word *scenario* is to emphasize the assumptions or hypotheses, and to de–emphasize predicting. The

TABLE 14-1.

Energy Requirements of Electric Motor Vehicles

Reference	Estimate	Conversion to Power Plant Ouput	Conversion to Power Plant Input
J.J. Gumbleton, et al (5)[a]	$0.195 \frac{KWH_w}{mile}$[b]	$0.214 \frac{KWH_e}{mile}$	$0.668 \frac{KWH_t}{mile}$
	$0.379 \frac{KWH_w}{mile}$[c]	0.417	1.305
P.D. Agarwal (6)	$0.200 \frac{KWH_p}{mile}$[d]	0.436	1.362
	$0.500 \frac{KWH_p}{mile}$[e]	1.090	3.415
B.C. Netschert (7)[f]	$0.362 \frac{KWH_p}{mile}$	0.790	2.470
Federal Power Commission (8)[g]	$0.500 \frac{KWH_w}{mile}$	0.549	1.717

a. Based on actual test of a specially built two-passenger battery-electric vehicle of weight 1,650 lbs.
b. Based on a constant 30 MPH speed over a distance of 47 miles.
c. Based on a "standard" two stop per mile driving schedule (15% time for acceleration, 17% for deceleration, 45% for constant speed, 23% for rolling.) Range 30.6 miles, average speed 14 MPH.
d. Assumes a "limited performance" vehicle of 2,000–2,500 lbs. with a range of 60 to 100 miles and a maximum speed of 50–60 MPH.
e. Assumes a standard sized car with air conditioning and power conveniences.
f. Assumes an average size passenger car with a weight of 4,000 lbs. and an average speed in use of 40 MPH.
g. Assumes a variety of conventional vehicles operating mainly in urban-suburban driving conditions.

scenarist does not predict or forecast. Rather he posits a set of assumptions or hypotheses, and on the basis of these assumptions moves to a possible view or scene of a future. With the scenarist thus defined, we are left with two questions. What is the value of a scenario? What rules should a scenarist follow?

The value of a scenario is that it suggests futures which may arise from certain actions or because of certain policies. It thus gives the planner or decision maker some information on which decisions to make. The scenario has two weaknesses. First, it may not lead to the anticipated future, even if the assumptions are followed, because the assumptions may be incomplete. Second, actions may not correspond to the assumptions made, and they may lead to an unexpected outcome. The alternative is non–planning, which historically leads to chaos. So the scenarist proceeds.

To be of interest a scenario should be credible or reasonable. One recent analysis of energy needs to power electric cars assumed an immediate change of all internal combustion vehicles to electric vehicles, and compared the new energy demand with the total national demand for that year. Such an assumption is clearly unrealistic because it obviously would require many years or decades to convert the automobile population to electric vehicles, if that were desirable. In the meantime, the electric power industry is growing rapidly. Any comparison of possible future demands should consider such factors.

In an attempt to establish a reasonable basis for a scenario we consider four reasons which could lead to the major development of the electric automobile.

1. It may be found to be less polluting than the present car.
2. It may require less energy than gasoline–powered automobiles require.
3. It may be found to be less expensive.
4. The available supply of petroleum may be exhausted.

Whether or not the electric automobile pollutes less than its internal combustion counterpart is very difficult to answer. The effect of the substitution would be an exchange of one form and magnitude of pollution for another form and magnitude of pollution. The present internal combustion automobile releases significant amounts of Hydrocarbons and Oxides of Nitrogen and Sulfur. It is a major contributor to urban smog. It does have the advantage of distributing its waste heat over a rather wide range. On the other hand, pollution related to the electric automobile comes primarily from the electric power plant, discussed in earlier chapters. There is no clear answer as to which pollution is more significant.(9, 10) It is the author's opinion that there is today no compelling argument for or against the electric automobile on pollution grounds.

A second argument for the electric car is that available petroleum resources may soon be depleted to such a degree that they can no longer be used for powering automobiles. The primary fuel for the electric automobile would probably then be coal or Uranium, at least for the next couple of decades. As we saw in Chapter 13, petroleum resources are very difficult to estimate, particularly because of the international character of this market. It does seem reasonable, however, that the demand for petroleum and its decreasing availability may force the cost of gasoline to rise, thus encouraging the development of the electric automobile. In our scenario below we assume that this pressure becomes significant in the next decade or so.

The cost of the electric automobile is also difficult to estimate. To date such vehicles have not been produced in large numbers. Projected costs suggest that electric cars may be at least as expensive as present cars and probably more so, largely because of the cost of batteries. There is no present evidence that costs will be a major factor in inducing a switch to the electric car. An important exception may be the substitution of a small low-power electric car for an unnecessarily large and powerful gasoline automobile for some purposes.

Finally, we come to the question of energy. In Section 14.5, we see that the primary fuel energy requirements of comparable electric and gasoline automobiles are quite similar. Hence, concern for total energy use does not appear to offer major motivation for the substitution of electric cars for gasoline cars.

On the basis of the above considerations, it appears that com-

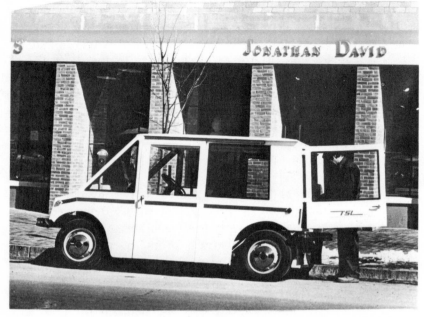

FIGURE 14—3. Prototype electric delivery van. The Anderson T/3 electric delivery van, now in preliminary production, is an example of a number of vehicles on the drawing boards or in initial production. This vehicle weighs 3,000 pounds, 900 pounds in the batteries alone. The initial cost of the vehicle is high—about $9,000—but operating and maintenance costs should be substantially lower than for equivalent conventional vehicles. Top speed is 40 mph. The range is about 30 miles for stop—and—go conditions. (*Courtesy Transportation Systems Laboratory, a division of Electromotion, Incorporated, Bedford, Mass.*)

pelling reasons do not exist today for a rapid shift to electric cars, unless such a shift implies lighter and lower–power vehicles. There is some evidence that substitution on a small scale may develop in the next few years, and substitution on a large scale could develop in the next 10 to 20 years.

Of course we must always keep in mind that basic assumptions or conditions which would alter the above conclusion may change. In a time of rapid technological and sociological evolution such changes could occur over very short periods of time. It is possible, for example, that major technological advances may make the electric automobile far more efficient than we assume here.

At this point we initiate our scenario by using the data of Table 14-1 to estimate the electric energy which might be required by hypothetical electric cars of the future. We consider two vehicles, one of "limited performance," and one "standard size" (analogous to today's average-size car). The assumed energies in KWH_e, KWH_t (and the equivalent BTU measures) are:

	KWH_e/mile	KWH_t/mile	BTU/mile
Limited Performance Vehicle	0.400	1.25	4,270
Standard Vehicle	0.800	2.50	8,540

The total annual power plant bus bar electric energy required by n_s standard vehicles driven an average of m_s miles per year, and n_ϱ limited performance vehicles driven m_ϱ miles per year is:

$$E_v = n_s m_s \ (0.8) + n_\varrho m_\varrho \ (0.4) \quad KWH_e \qquad (14\text{-}4)$$

The reader can of course substitute whatever values he considers reasonable for the parameters in Equation 14-4 to find the electric energy required for a given year. As examples we consider two scenarios.

Scenario 1. On the basis of the number of automobiles in the United States in 1960 and 1970(11) we hypothesize a 1985 automobile population of 125 million. It is assumed here that this includes 5 million "limited performance" electric vehicles, and that their average annual mileage is 4,000 miles. Another 5 million "standard" electric vehicles are assumed, with an average annual mileage of 12,000 miles. We further assume that the electric energy growth rate continues at about 7 percent per year for a 1985 United States consumption level of 3,900 billion KWH_e. The power plant energy required to power the electric automobiles assumed for 1985 can be obtained from Equation (14-1) as:

$$E_v = (5 \times 10^6) \ (1.2 \times 10^4) \ (0.8) + (5 \times 10^6) \ (0.4 \times 10^4) \ (0.4)$$

$$= 56 \times 10^9 \ KWH_e$$

This represents 1.4 percent of the assumed 1985 electric energy demand. That is, while electric energy demand is increasing by 180 percent from all causes during the 15 years from 1970 to 1985, the increase due to a rather modest fleet of 10 million electric automobiles, represents 1.4 percent of the total demand in 1985.

Scenario 2. In this scenario we assume that a much larger fleet of electric vehicles develops by the year 1985. Assume 40 million standard vehicles and 60 million limited performance vehicles (same mileage as the first scenario in both cases).

$$E_v = (4 \times 10^7)\,(1.2 \times 10^4)\,(0.8) + (6 \times 10^7)\,(0.4 \times 10^4)\,(0.4)$$

$$= 480 \times 10^9 \text{ KWH}_e$$

In this scenario electric automobiles require 12.3 percent of the 1985 electric energy. Some writers feel that the assumption of a

FIGURE 14–4. Small electric automobiles. Light electric automobiles such as these require far less primary fuel energy than the average 3,500 to 4,000 pound gasoline automobile uses. The size, cost, and energy–storage capacity of batteries may limit development of these vehicles. Better batteries and more efficient power systems could lead to major development of the electric car. (*Courtesy Pacific Gas and Electric Co.*)

continuing 7 percent growth is unrealistic. The implication of the scenario above is that development of a large electric automobile fleet would have a tendency to maintain a high growth rate.

The above scenarios, particularly the second, imply that the electric automobile may become an important factor in the energy crisis. It is not clear, however, in what way the electric automobile will affect the problem. It will, of course, lead to a need for more electric power plants and fuels for these plants. On the other hand, it will reduce the demand for gasoline (and therefore petroleum) as a fuel source for automobiles, as long as the electricity is generated by some other fuel, such as Uranium, coal, natural gas, or water power. If a plentiful fuel, such as coal or Uranium (in a breeder reactor) is used cleanly and safely, the electric automobile may in some ways be an attractive alternative to the gas–burning and polluting, internal–combustion automobile. We shall discuss this question more philosophically below. But first we turn our attention briefly to a comparison of the energy requirements of electric and gasoline automobiles.

14.5 COMPARING ELECTRIC AND GASOLINE AUTOMOBILES

If we assume a power plant efficiency of 32 percent, the limited performance vehicle requires a power plant thermal energy input of 4,270 BTU/mile. The standard vehicle requires 8,540 BTU/mile.

To determine the energy required by the gasoline automobile, assume the energy available from gasoline to be 136,000 BTU per gallon. According to a recent study the average consumption rate of gasoline for a standard size automobile is 13.6 miles per gallon, and for a "subcompact" size automobile, 21.43 miles per gallon.(12) We assume here that the subcompact automobile is somewhat larger and more powerful than the limited performance electric vehicle discussed earlier. For comparison purposes we assume a limited performance gasoline automobile with consumption rate of 28 miles per gallon.

The table below summarizes the fuel energy requirements of gasoline and electric automobiles.

	Electric	*Gasoline**
Limited performance	$4,270 \dfrac{BTU}{Mile}$	$4,860 \dfrac{BTU}{Mile}$
Standard size	8,540	10,000

In considering the above table the reader should recall that a number of assumptions are made in reaching each of these four numbers. The inaccuracies associated with these assumptions may lead to total errors of as much as 20 percent. If the errors are no greater than this, we can make the following very important general conclusion: There appears to be little difference in the primary fuel energy required to propel comparable electric and gasoline automobiles.

The conclusion above is not too surprising, since comparable vehicles have about the same amount of *work* to do in providing transportation, and both employ relatively inefficient power plants.

A public system which is somewhat comparable in personal convenience and privacy to the automobile is Personal Rapid Transit (PRT). In such systems small vehicles on guideways are dispatched on customer call to provide service to one or a few riders. Very little data is as yet available on energy costs of such vehicles. One source suggests a power consumption per seat of two to five times that of a large rapid transit vehicle.(13) This should put its energy cost per vehicle mile close to that of the limited performance vehicle. Hence, this system does not appear to offer an improvement in energy efficiency over the automobile. Again, this is not surprising, since a PRT vehicle which is comparable in weight to an electric automobile should have about the same work to do in accelerating and overcoming grades, friction, and air resistance. (We have not, of course, discussed the indirect energy costs of

*These figures do not take into account the effect of 1976 automobile emission standards on fuel consumption, which will almost certainly increase the required BTU/mile.

PRT and the automobile.) The advantages of PRT are that vehicles are shared, so that it would not be necessary to build a great many to serve a population; also, PRT does not require highways. The disadvantage of PRT is that it requires specially–built guideways.

All the comparisons above have been on the basis of energy cost per vehicle mile. If a vehicle or system carries more than one passenger, the cost per passenger mile decreases. Since this is the more significant basis of cost, the degree of loading is very important in comparing systems. The effects of loading are discussed elsewhere.(14) Here we note that there is no particular reason to expect loading of electric automobiles to exceed that of gasoline

FIGURE 14–5. The PPV, or People Powered Vehicle, is a lightweight three–wheeled two–seater with room for special cargo. Engineered for ease and safety of operation, the device features a three–speed transmission. It is usually licensed as a bicycle. A battery-powered version is under development. (*Courtesy EVI, Inc.*)

automobiles. PRT systems are too many in form to discuss in general.

We conclude that all three of these systems, when operated at about the same level of comfort, privacy and convenience, appear to require about the same amount of fuel energy.

14.6 SOME COMMENTS IN CONCLUSION

The objective of this chapter has been to consider the effects which the advent of the electric automobile might have on the "energy crisis." We are left with two important general conclusions.

1. There appear to be few major differences in the propulsion energy requirements of the gasoline automobile, the electric automobile, and (probably) PRT, where they provide comparable service.
2. Justification for the electric automobile, given the present technology, apparently must come from considerations other than energy cost, such as decreased pollution or a substitution of an alternate fuel if gasoline becomes very scarce.

Electric automobiles are nothing new. They were an important factor in the early development of the automobile. But the internal combustion car won out, largely because of the size and cost of storage batteries used in electric cars. This problem has not been solved, but there is optimism today that important advances can be made in electric batteries in the near future.

The problems posed by the gasoline automobile, as a major consumer of resources and source of air pollution, are increasingly apparent. As we noted in Chapter 13, one energy study in California in 1973 proposed a heavy tax on vehicles over 1,600 pounds in weight. If we do build electric vehicles in large numbers, it seems apparent that they should be light–weight vehicles which would conserve substantial amounts of energy. Converting from a large fleet of heavy and high–powered gasoline automobiles to smaller electric vehicles would have major effects on the energy problems outlined in Chapter 13. First, there would be a major reduction in petroleum demand. Second, the primary source of energy for the car would be diversified. Cars would be powered, indirectly, by coal, gas, Uranium, the sun, or any other source of electric energy.

Beyond the automobile, gasoline or electric, are a number of alternatives which are much less energy–intensive. To these we might turn—or return:

Mass transit   Mass transit, whether electric or gasoline–powered, when well loaded, can have an energy advantage of about five to 15 times over the automobile.(14)

Bicycles   A bicycle requires around 130 BTU/mile.(15) This is an advantage of roughly 25 times over the small vehicle.

Walking   R.A. Rice estimates that a walker uses about 400 BTU/mile. The advantages, perhaps, go far beyond the energy gain.(15)

REFERENCES FOR CHAPTER 14

1. E. Hirst and R. Henredeen, "Total Energy Demand for Automobiles," International Automotive Engineering Congress, Detroit, Michigan, January 8–12, 1973.
2. Data obtained from various *Consumer Reports* automobile surveys.
3. W.D. Ruckelshaus, speech at the Highway Research Board, Washington, D.C., January 24, 1973, available from Environmental Protection Agency.
4. "Control of Air Pollution from New Motor Vehicles and Engines—Federal Certification Test Results for 1973 Model Year," Environmental Protection Agency, Washington, D.C., May 2, 1973.

5. J.J. Gumbleton, D.L. Frank, S.L. Genslak, A.G. Lucas, "Special Purpose Urban Cars," Society of Automotive Engineers, Mid-Year Meeting, Chicago, May 19-23, 1969.

6. P.D. Agarwal, "Electricity Not Such a Clean Fuel," *Automotive Engineering*, February, 1971, pp. 38-39.

7. B.C. Netschert, "The Economic Impact of Electric Vehicles: A Scenario," *Bulletin of the Atomic Scientists*, May, 1970, pp. 29-35.

8. Federal Power Commission, "Development of Electrically Powered Vehicles," 1967.

9. S.K. Mencher, H.M. Ellis, *The Comparative Environmental Impact in 1980 of Gasoline-Powered Motor Vehicles Versus Electric-Powered Motor Vehicles*, Gordian Associates, Inc., New York, October, 1971.

10. J.T. Salihi, "The Electric Car—Fact and Fancy," *IEEE Spectrum*, June, 1972, pp. 44-48.

11. *1971 Automobile Facts and Figures*, Automobile Manufacturers Association, Inc., Detroit, Michigan.

12. *Cost of Operating an Automobile*, U.S. Department of Transportation, U.S. Government Printing Office, April, 1972.

13. A.J. Sobey, "Economic Considerations of Personal Rapid Transit," *Personal Rapid Transit*, Institute of Technology, University of Minnesota, April, 1972, pp. 207-230.

14. E. Hirst and T.J. Healy, "Electric Energy Requirements for Environmental Protection," National Science Foundation Conference on Energy: Demand, Conservation and Institutional Problems, Cambridge, Mass., February 12-14, 1973.

15. R.A. Rice, "System Energy and Future Transportation; Energy Technology to the Year 2000," *Technology Review*, Cambridge, Mass., 1972.

PROBLEMS FOR CHAPTER 14

General Problems

14.1. About how many barrels of petroleum were used in 1972 to power gasoline automobiles? How many barrels is this per person, assuming a population of 210,000,000? (1,650 million barrels; 7.8 barrels per person)

14.2. If the amount of petroleum used in 1972 for automobile fuel had been cut in half by various conservation measures, what would be the savings in petroleum in barrels per day? (2.2 million barrels per day)

14.3. About how many BTU's per mile are required to propel a 2,000 pound vehicle? A 5,000 pound vehicle? (5,000; 12,500)

14.4. What is the approximate gasoline mileage to be expected from a 1,900 pound vehicle such as a small Volkswagen? (28.7 mpg)

14.5. What is the approximate mileage of a vehicle with an engine displacement of 400 cubic inches? (10 mpg)

14.6. Using the assumptions in this chapter, find the mechanical propulsion energy, KWH_p, if the electric energy in the battery is 70 KWH_b. (39.2 KWH_p)

14.7. If a propulsion energy of 50 KWH_p is required, what is the necessary power plant fuel energy in KWH_t and in BTU's? (340 KWH_t; 1,160,000 BTU)

14.8. Find the total annual power plant bus bar electric energy required by 100,000,000 limited–performance vehicles driving an average of 8,000 miles per year. What percentage is this of the projected demand in 1980, considering an expected 7% increase per year?

14.9. Why do you suppose bicycling requires less energy than walking does?

Advanced Mathematical Problems

14.10. Derive Equation (14–2) from (14–1), showing how the units change.

14.11. Determine the relation between weight W and displacement D implied by the approximate Equations (14–2) and (14–3). (This is an average relation, and should not be expected to hold for any specific vehicle.) Plot W against D.

14.12. Construct a 7 by 7 square matrix of multiplying factors which converts any given energy form in Figure 14-2 to any other energy form. (Use the efficiencies assumed in Figure 14-2.)

14.13. Write a new equation corresponding to Equation (14-4) under the assumptions that the battery discharging efficiency increases to 90% and the electric control and motor efficiency increases to 85%. What is the percentage in total energy savings with these new efficiencies, compared with the lower values? (27%)

Advanced Study Problems

14.14. Write a paper which discusses in as much detail as possible the likely effects on society of replacing 125,000,000 standard–size gasoline cars with 125,000,000 limited-performance electric cars.

14.15. Find out how energy delivered to an automobile from gasoline or from an electric battery is used. What are the losses? How do they vary with speed, weight, and other factors?

14.16. In your opinion, would it be desirable to switch from gasoline to electric cars? Explain your answer in detail.

14.17. Explain in what ways pollution would change if we replaced gasoline cars with electric cars.

15

THE ELECTRIC POWER INDUSTRY

The worth of a state, in the long run,
is the worth of the individuals composing it.

John Stuart Mill
Liberty

The electric power industry is made up of a large number of private companies and public agencies which perform one or more of the functions of generating, transmitting, and distributing electric power. Closely associated with these companies or agencies are the builders or vendors of equipment required for these functions and the fuel suppliers who provide primary energy sources. We will also discuss in this chapter the public regulatory agencies which control the activities of the industry. The relation of these various bodies is summarized in Figure 15-1.

15.1 SEGMENTS OF THE ELECTRIC POWER INDUSTRY

There are four kinds of organizations in the electric power industry. The first is the investor–owned (private) power company. The second is the Federal agency, such as the Bonneville Power Administration, or the Tennessee Valley Authority. The third is the non–Federal agency, such as any municipal utility. The fourth is the cooperative established under the Rural Electrification Administration. Table 15–1 shows the number of each type of agency, and the percentage of the total energy produced in the United States which was generated by each in 1970.

Although investor–owned companies are few in number (about 12 percent of the total), they generate over three–fourths of the energy in the nation. Most serve the three–fold role of generation, transmission, and distribution. Such systems are said to be *vertically integrated.* Investor–owned systems are usually granted exclusive franchises from state or local government agencies, entitling and obligating them to provide adequate service to all customers in a specified area. In turn they are usually tightly regulated by some form of public utilities commission. (See Section 15.3.)

There are five Federal agencies: the Tennessee Valley Authority (TVA); the Bonneville Power Administration (BPA); the Southwestern Power Administration (SWPA); the Southeastern Power Administration (SEPA), and the Bureau of Reclamation. TVA and BPA were originally established during the 1930's to develop the hydroelectric potential of the Tennessee River and the Columbia River Basin. In more recent years TVA has added fossil–fueled and nuclear generating plants. BPA markets hydroelectric power gen-

TABLE 15-1.

Agencies in the U.S. Electric Power Industry (1970)

Industry Segment	Number of Agencies	Percent of Generation
Investor–owned	400	77.3
Federal	5	12.2
Public (non–Federal)	2,060	9.1
Cooperative	955	1.4

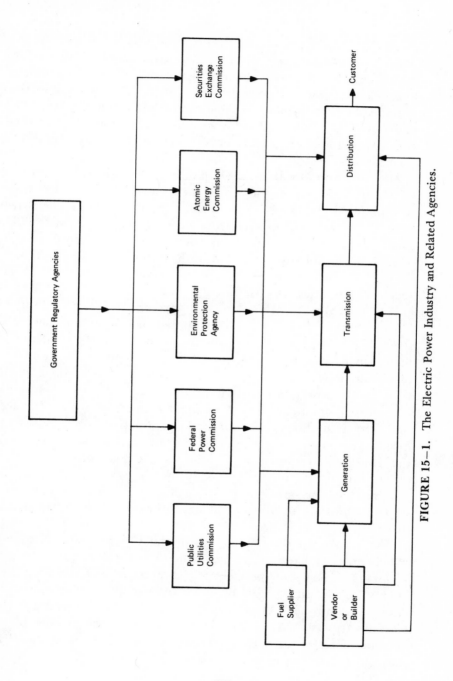

FIGURE 15–1. The Electric Power Industry and Related Agencies.

327

erated by 33 projects of the Bureau of Reclamation and the U.S. Army Corps of Engineers. It also owns very extensive transmission networks. The remaining Federal agencies generate and transmit hydroelectric power in their respective areas. With the exception of power sold to a number of large industrial customers, most Federal power is wholesaled to other segments of the industry. Federal law requires that preference be given to public agencies in such sales.

Most of the more than 2,000 public (non-Federal) agencies are owned by cities ranging in size from millions of people down to a few hundred. Only about 35 percent (700) generate their own power. The rest buy power from other generating agencies. The small percentage which generate their own power is indicative of one of the problems associated with having many small utilities. As we have seen earlier, there are important economy of scale effects in generating electric energy. That is, the cost per KWH of energy goes down as the plant size goes up. Hence, a small plant built to serve a relatively small number of people will have relatively high energy costs. A second problem is that the existence of many small systems in a region complicates regional coordination of power flows; such coordination is important to the dependable delivery of power where it is needed. There are two solutions. One is to combine a number of small systems into one. The second is to strengthen agreements and coordination between systems in a region. As a general rule the various agencies and systems in the United States cooperate quite well. Unlike private utilities, public utilities do not pay Federal income tax, and generally do not pay local or state taxes, although they often make payments or contributions to their local governments in one form or another.

The nearly 1,000 cooperatives in the United States also vary widely in size, from as few as a hundred to as many as 35,000 members. A very small percentage (6.8 percent) generate their own energy. Cooperatives are financed by the Rural Electrification Administration (REA), established in 1935, to encourage the electrification of farm areas through loans which are available at very low interest rates. Cooperatives also pay no Federal taxes but may pay local taxes in some regions.

Elements of the electric power industry are by no means isolated. The United States has a very extensive interconnection of facilities. Some transmission line ties are strong, some weak. The major advantage of interconnections is to allow systems to sell or trade power when it is surplus in one region and in short supply in another. The United States and part of southern Canada are broken into nine regions or areas where systems are tied together. The entire network is then coordinated by the National Electric Reliability Council.(1)

15.2 SITING ELECTRIC POWER PLANTS

One of the most important problems related to the growth of the electric power industry is the *siting* or location of new power plants.(2) This is usually an individual utility or government agency decision, though it often has important regional implications which must be considered.

The term "siting" can refer to the rather narrow process of deciding where to put an electric power plant. In this section, however, we use the term in reference to the very complex and lengthy process which starts with the realization that additional power is needed, and ends with choice of a type of plant, a location, and an analysis of benefits and costs. A number of possible sets of factors for consideration might be specified. The following list is rather arbitrary but fairly complete for our purposes. These are essentially decisions which must be made by the utility.

1) Plant size
2) Development schedule
3) Type of plant
4) Plant location
5) Location politics
6) Environmental effects
7) Participation by other utilities
8) Licensing
9) Construction
10) Operation
11) Costs

These factors are not necessarily completed one-by-one in this order. It is often necessary to come back and review one preliminary decision in the light of later considerations.

Inasmuch as this is perhaps the basic growth decision of a company, it is not surprising that the decision process involved makes use of most of what has been discussed in this book.

15.3 MARKETING ELECTRIC POWER

The electric power industry, like most industries, has for decades vigorously sold its product, encouraging increased use of electricity. In the early years of the industry this was perhaps important and desirable. It educated the public to the availability of a versatile form of power which was clean at the point of use. In recent years, however, there has been increasing opposition to the promotion of electric power.(3) Most companies have either eliminated or radically changed their promotional policies.(4, 5) Two factors have led to this change. The first is environmental protection pressure from persons and organizations arguing against the selling of additional energy. The second is the very real shortage of electric energy which is either experienced or anticipated by many utilities.

Many industries are moving in the direction of promoting energy conservation. In 1971 the Consolidated Edison Company in New York initiated its "Save-a-Watt" program, urging its customers to turn off air conditioners, lights, and appliances wherever possible. The program had some success and helped New York get through a hot summer without serious power curtailment.

Although the electric power industry may at some times promote use, and at other times discourage it, it is not free to dictate use. As Charles Luce, Chairman of Consolidated Edison, put it, "I don't think people should be free to use all the electricity they are willing and able to pay for, if that implies waste. Electricity and all other forms of energy must be used wisely and not wastefully if the earth is to remain a good place to live. But utilities cannot tell people they cannot have all the electricity they want, or decide what is wise use and what is wasteful use. That is a decision each individual must make for himself—or, possibly, that our elected officials will be forced to make for all of us."(6)

While the power industry cannot decide what is wise use and what is wasteful use, it can help citizens and elected officials make that decision by providing open and unbiased information. This educational role, which would serve the public interest in general, is perhaps the most desirable form for a new marketing task for the industry.

15.4 BUILDERS AND VENDORS

Closely associated with the electric power industry are those companies and agencies which provide materials necessary for the operation of the industry. These can be classed as:

1) Construction companies
2) Equipment builders and suppliers
3) Fuel suppliers

Construction companies can be huge organizations necessary to build a large dam or major power plant, or they can be small companies building a sub-station or a short power line. Equipment companies build and supply everything from huge 1,000 MW generators to tiny system components. These first two groups are basically responsible for the capital or fixed costs of electric energy. Fuel suppliers can provide coal, oil, natural gas, nuclear fuel, or geothermal steam. Suppliers are usually private companies, although nuclear fuel has in the past been supplied by the Federal Government through the Atomic Energy Commission. In the future, private companies will produce nuclear fuels.

The electric power industry and its suppliers maintain close contacts for the purpose of exchanging information, setting standards, and developing new systems.

15.5 REGULATORY AGENCIES

A number of state and Federal agencies control the operations
of the electric power industry. These agencies have been established for two basic reasons. The first is to control the direct relations of the industry with the public, primarily in energy pricing,
because most power companies are in a monopolistic position. The
second reason is to control indirect effects on the public and on
other industries, relating to the control of water use, the use of
radioactive materials, and the protection of the environment. The
agencies which have perhaps the greatest effect on the industry
are:

1) Federal Power Commission (FPC)
2) Public Utilities Commission (PUC)
3) Atomic Energy Commission (AEC)
4) Environmental Protection Agency (EPA)
5) Securities and Exchange Commission (SEC)

The Federal Power Commission is the major Federal agency responsible for assuring that the United States has an adequate supply of electric power available at the most economical prices
possible.(7) It is also charged with assuring proper use and conservation of natural resources. Besides these general responsibilities
for the health of the industry, the Commission has a number of
specific responsibilities. It licenses most non–Federal hydroelectric
facilities, establishes accounting practices, regulates interstate rates
and activities, and controls some aspects of private utility activities.

Public Utilities Commissions, often with some other formal
name, exist in 46 of the 50 states.(8) About half control both
public and private utilities. The other half control only private or
investor-owned utilities, assuming that public utilities do not require control. These state commissions have a very wide range of
responsibilities in the various states. Typically, they set rate structures at a level adequate to insure a fair and reasonable profit to
private investors. They establish accounting procedures, assure
adequacy of supply and safety, and certify expansion or development of systems under their control.

FIGURE 15−2. A regulatory hearing. The electric power industry is regulated by a wide variety of public agencies. Hearings before such agencies may concern a raise in electric energy rates or the construction of a major new power plant. The utility petitioning for the proposed action must show the desirability of its plans. Hearings also provide other interested companies, groups, or the general public with the opportunity to comment on the proposed action. (*Courtesy Pacific Gas and Electric Co.*)

The Atomic Energy Commission regulates construction and operation of all nuclear reactors in all segments of the industry. Regulation extends to the fuel cycle, the design of the plant, including safety features, and the disposal of waste products. The AEC has an elaborate licensing procedure which starts with the initial proposal of a utility to build a plant, extends in various steps through construction, and finally into the operation of the plant. Licensing is intended to assure protection to the public from radiation hazards, and protection to the environment. Public hearings are held at various stages of the process. The elaborate licensing procedure, plus the construction complexities of a nuclear power plant, have combined to cause a delay time from first conception of a plant to initial operation of about 10 years. This

long lead time has a major effect on the growth of the electric
power industry. In recent years there has been some controversy
over the dual role of the AEC as both promoter and regulator of
nuclear power. In early 1972 the AEC announced that it would no
longer play a promotional role.

The Environmental Protection Agency was established to carry
out the mandate of the National Environmental Protection Act
(NEPA) of 1969. The most important section of NEPA, as far as
the electric power industry is concerned, is Section 102, which
requires the preparation of an environmental impact statement in
connection with any major Federal project which would have a
significant effect on the environment. Section 102 also requires
discussion of all alternatives to the proposed action, and discussion
of any irreversibility in the action. The courts have been strict in
their interpretation of NEPA, and it appears that it will have a
significant effect on the environment. It has also had the effect of
increasing the time required to put a plant on line, and of restricting the siting options of the power company. It will almost certainly lead to greater electrical cost, since pollution control costs
must be passed on to the consumer.

Finally, the Securities and Exchange Commission regulates the
structure and activities of holding companies, the form of mergers
of companies, issuance of securities, and various other economic
and business practices of the companies in the industry. Its role
in the future development of the industry is important because of
the movement toward merging of companies and intercompany
transactions.

One of the most difficult problems relating to regulatory agencies arises from the backgrounds of their members. In many cases
it is either desirable or necessary that a number of members of
an agency or commission come from the industry which is being
regulated. It is often, though not always, difficult to find persons
outside an industry with sufficient knowledge or expertise to be
effective regulators. Commissioners coming from an industry may
bring with them a level of interest in or relation to the industry
which makes completely impartial decisions difficult.

15.6 RESEARCH AND DEVELOPMENT

All industries must devote a certain amount of money to re-search. The amount of money spent is highly dependent on the nature of the industry, the degree of competition, and the perception of the problems facing the industry. For many years prior to about 1970, the electric power industry was perceived by many to be relatively static. Its methods were established: meeting demand was largely a question of building new plants much like the old plants. This is not to suggest that no progress took place. In fact, some extraordinarily important developments occurred in the 30 years prior to 1970 in such areas as fossil–fuel plant efficiency, nuclear power, and power transmission. Nonetheless, the changes were gradual and the percentage of revenue devoted to research expenditures was small compared with more volatile segments of the technological community. In the latter part of the 1960's a new perception of the problems faced by the industry evolved, and with it came a new interest in research and development.

Expenditures for research and development in the 1960's averaged nearly one fourth of one percent of gross revenues for the utilities. In 1969 the utilities spent about \$40 million. Equipment manufacturers spent an additional \$110 million. It seems very likely that utility and government expenditures will rise substantially, the former by perhaps three to four times. Various proposals have been made to tax electric energy for research. If each KWH produced in the United States were taxed at a rate of 1/4 of a mill, about \$400 million would be raised each year.

There is substantial disagreement on many aspects of future research and development. One is the form of financing. A second is the form of the body which would direct research. It seems to be fairly clear that the main research interests of the electric power industry center around how electric energy is generated and distributed, rather than how it is used. It is an essential position of the industry that its responsibility is to meet demand, and not to question the nature of the demand. This appears to leave to other groups the question of efficiency of use, and perhaps even pro-

FIGURE 15−3. Electricity for farming. The flow of electricity to rural areas
has made available the conveniences and economic opportunities that were
once found only in the city. In the past 40 years, various government programs such as the Rural Electrification Administration and the five major
Federal Regional Administrations have brought electricity to the farm. The
dairy farm above receives power from the Tennessee Valley Authority. When
TVA started, only three out of a hundred farms in the valley had electricity.
Now practically all are electrified. (*Courtesy the Tennessee Valley Authority.*)

priety of use. Whether the latter is anybody's business is a question
of much controversy. We shall consider that question in the next
chapter.

On January 1, 1973 the Electric Power Research Institute
(EPRI) came into being. EPRI was created to conduct a coordinated research and development program, with support from all
segments of the power industry. It is a major new effort on the
part of the industry to meet the technological challenges of the
years ahead.

15.7 SOME OBSERVATIONS ON THE ELECTRIC POWER INDUSTRY

For years the electric power industry played a highly successful role (according to its charter of the time), largely out of the public eye. It was simply assumed that the industry was doing its job. But in the late 1960's the rules of the game changed for a number of reasons. Costs rose sharply, fuel availability began to be ques-

FIGURE 15–4. The building of the electric power industry. In the first decade of the twentieth century, 14 horses pulled part of an electric generator past a quiet river in the Sierra Nevada in California. When the stream was finally tamed and harnessed, it provided the power of 100,000 horses to the growing state of California. Man had conquered nature. In doing so, he had gained something, and he had also lost something. He weighed costs and benefits and made his decision. He can never do less. The rules may change. Perspectives may also change. But every decision must be made. (*Courtesy of Southern California Edison Co.*)

tioned, and perhaps most significant of all, the American people developed a new interest in the environment. The electric power industry was suddenly thrust into the public eye. Its sins were denounced, its problems laid bare. Every industry has both, but one so long unused to public scrutiny is particularly unready for such analysis. The natural result was a posture of defense, with an increasingly critical analysis in response. Fortunately, there is some evidence that both the industry and its antagonists are beginning to listen to and understand each other's viewpoints. It is not suggested here that there should ever be complete agreement on the controversies over power. It is, however, very necessary that there be general agreement on two points. First, the life which most Americans lead and desire depends in part on the availability of some electric energy. Second, no commodity or product, including electric energy, can be developed independently of consideration of its impact on society; and, such consideration will moderate development. The electric power industry will no doubt be called upon to play an increasing role in effecting rational compromises between demand and impact.

REFERENCES FOR CHAPTER 15

1. *1971 Annual Report—National Electric Reliability Council and its Regional Councils*, Research Park, New Jersey, April, 1972.
2. *Engineering for the Resolution of the Energy–Environment Dilemma*, National Academy of Engineering, Washington, D.C., 1972.
3. E. Hirst, "Debunking Madison Avenue," *Environmental Action*, February 10, 1972, pp. 8–9.
4. T.J. Galligan, Jr., "The Future Role of Marketing in the Electric Utility Industry," *Public Utilities Fortnightly*, May 10, 1973, pp. 42–44.
5. W.A. Knoke, "Changing Marketing Practices in the Electric Power Industry," *Public Utilities Fortnightly*, May 10, 1973, pp. 45–61.
6. C.F. Luce, "Can People Have the Electricity They Need?" Speech before the Supplemental Training Program, AT&T, New York City, October 13, 1971.

7. *Federal Regulation of the Electric Power Industry Under Parts II and III of the Federal Power Act*, Federal Power Commission, Washington, D.C., July 1, 1967.
8. *Federal and State Commission Jurisdiction and Regulation of Electric, Gas, and Telephone Utilities*, Federal Power Commission, Washington, D.C., 1967.

GENERAL READING FOR CHAPTER 15

1. E. Vennard, *The Electric Power Business*, Second Edition, McGraw-Hill, New York, 1970.

   This is an excellent, easily-read overview of the industry, written from the perspective of the industry. Covered are such topics as regulation, economics, forecasting, marketing, power loads, costs of power, and relations with the community.
2. *The 1970 National Power Survey (Part I)*, Federal Power Commission, U.S. Government Printing Office, December, 1971.

   The National Power Surveys cover the entire industry periodically. Most of this large report is quite readable. It also covers most of the topics covered here.
3. Reference 2 above.

   This is an excellent general review of a number of technical topics, presented on a relatively simple level. It is largely oriented toward siting decisions.
4. P. Sporn, *The Social Organization of Electric Power Supply in Modern Societies*, M.I.T. Press, Cambridge, Mass., 1971.

   This is a very interesting non-technical discussion of power in different social-political environments. It is written by one of the industry's foremost spokesmen.

PROBLEMS FOR CHAPTER 15

General Study Problems

15.1. Identify the type of utility which delivers power to your community. What is the source of that power?

15.2. Consider that the utility which services your home determines that it must add a plant ten years from today that is 10% of the present capacity. Suggest the plant type and location which you feel appropriate. Consider as many of the planner's problems as you can. Assume a base load plant.

15.3. Determine some of the major problems which require research in the electric power industry.

15.4. If possible attend a hearing of a regulatory agency on a matter relating to electric energy.

16

THE DILEMMA AND
THE CHALLENGE

Tyger, tyger, burning bright
In the forest of the night.
What immortal hand or eye
Could form thy fearful symmetry?

William Blake

In this book we have posed a dilemma, and we are left in the end with a challenge. The dilemma is the difference between the rapidly growing demand for energy and the supply of available energy. The challenge is to find a way, acceptable to society, to eliminate the difference between supply and demand.

In this chapter we shall be discussing energy in general, as well as electric energy in particular. An important objective of Chapters 13 and 14 was to show how inseparable these subjects are. We began this book with an emphasis on electric energy because it is the fastest-growing form of energy, and because it is an excellent example for the study of energy problems in other fields. As we have moved along we have built up to the clear statement that we cannot speak of an "electric energy crisis" in isolation from an "energy crisis."

16.1 THE DILEMMA OF SUPPLY AND DEMAND

This book has presented the problem of demand and supply in some detail. We need not review at this point that problem at any length. In this section we shall very briefly summarize the problem.

Electric energy demand is growing at a rate of about seven percent per year, with a doubling time of 10 years. Total energy use grows at about three and a half percent per year, with a doubling time of 20 years.

Supplying electric energy is becoming increasingly difficult with diminishing fuel reserves, environmental opposition, and increased power plant construction times. A number of new and largely untested systems may help meet future demand. It is generally, however, difficult to estimate the likely success of these approaches.

Supplying total energy is a matter of finding and developing *available* energy resources. Some of these, such as solar power and hydropower, are non-depletable. Others, such as the fossil fuels and Uranium, are depletable. Eventually we must move toward entire dependence on non-depletable resources. In the meantime we shall continue our major reliance on depletable fuels, primarily the fossil fuels, for the next one to two decades at least.

16.2 THE CHALLENGE

The problem is that in many areas energy demand either exceeds supply today, or is expected to exceed supply in the near future. In principle the solution to the problem is simple. We must either increase the supply or decrease the demand. In practice we shall find it very difficult to do either without disturbing major elements of our society. Let us consider the problem of increasing supply.

Many studies today center almost entirely upon the problem of meeting demand as it appears.(1, 2, 3) Most of the power industry's work is directed toward meeting demand, partially because that is its mandate, and partially because this is its tradition—not easy to break. The industry's efforts have largely, though not entirely, followed traditional development lines, working on approaches which were essentially proven. It has been left largely to government to seek exotic new sources such as fusion and solar power. The industry, supported by the present administration (1973), emphasizes fossil–fuel reserves and development of the breeder fission reactor.(4)

A bill introduced recently in Congress seeks less certain solutions.(5) It calls for a ". . . development strategy designed to make available to American consumers our large reserves of domestic fossil fuels, nuclear fuels, and geothermal fuels, and the potentially unlimited reserves of solar power, nuclear, and other unconventional sources of energy." Specifically, the bill would establish five corporations for the purpose of developing coal gassification; shale oil resources; advanced power cycles; geothermal energy, and coal liquefaction.

On June 29, 1973, President Nixon announced several moves to counter the apparent threat of energy depletion. He appointed Governor John A. Love of Colorado Director of the Energy Policy Office, an agency of the Executive Branch which Mr. Nixon had begun under another name just two months previously. He also requested Congressional authorization for funding for the Energy Research and Development Administration, an arm of the Federal Government which would explore the most effective means of producing energy—especially energy from coal.

In addition, Mr. Nixon asked the American public to drive smaller cars and to moderate both the heating and cooling requirements of their homes in order to conserve energy.

If we cannot or do not wish to increase supply sufficient to meet demand, then we must reduce demand. We must conserve energy. We may do this voluntarily, or it may be forced on us as in the case of a gasoline shortage or a power blackout, or it could be legislated by Congress or a state legislature.

We arbitrarily divide energy conservation into three categories:

1) Conservation that people find generally desirable
2) Conservation that is somewhat undesirable, leading to minor inconveniences
3) Conservation that many people find highly undesirable, with a major impact on life style

Individuals will not, of course, agree generally with specific examples in these categories, but it is of value to know that there are degrees of conservation.

In the first category we find a number of areas in which the consumer can save money and energy at the same time without significant inconvenience or loss of desired amenities. Major savings in energy come from insulating homes and weatherstripping doors to prevent unwanted air flow into a heated or cooled house. The original capital cost may appear significant, but it is usually paid off through reduced operating costs in a few years. Some major appliances, such as air conditioners, are built in different models with different energy efficiencies, but little or no difference in capital cost or performance. Careful, educated shopping can save the consumer operating costs and the country energy. Energy costs for operating large and small buildings can be significantly reduced by improved building design.(6) Energy costs the consumer money, and it is very reasonable that he take advantage of any savings in energy use which does not change his life style in an important way.

Beyond the easy changes that can save energy and money without hurting, we look to ways of saving that may be somewhat inconvenient or undesirable. If the energy crisis increases, it may become increasingly important for the consumer to make certain relatively minor sacrifices. There are many ways to cut energy use. Turn off lights not actually in use. Turn off the air conditioner when you leave the house for a number of hours. Use

FIGURE 16–1. Four Corners coal–fired plant. This is one of the fossil–fuel electric power plants built in the Four Corners area of Utah, Colorado, New Mexico, and Arizona. Air pollution is a growing problem here despite the use of electrostatic precipitators (visible on the near side of the plant). Coal–storage piles are visible on the right. A cooling pond is behind the plant. (*Courtesy Southern California Edison Co.*)

public transportation, a bicycle, your legs. Avoid buying or using unnecessary appliances. Drive a lightweight car. Use returnable bottles instead of cans for beverages. Organize car pools. Hang your clothes out to dry in the sun. All of these mean inconvenience for some people; they also mean energy savings and dollar savings.

If voluntary savings are not enough, and if demand grows faster than supply, we may someday face major government control of energy use. Gasoline rationing, restrictions of car size, cutting off power to some industries, and severe building restrictions are all possible if man does not effect a balance between demand and supply. The voluntary approach seems desirable.

16.3   THE PROBLEM OF VALUES

In attempting to determine the best course of future action it is useful to have guidelines, principles, or values to help direct activity. But we have few, if any, generally accepted values which might guide us to a "best" solution. We have found it difficult to articulate a value structure for weighing the costs and benefits to society of technological options.

The debate over technology and environmental balance is strongly polarized today. On one hand some argue that we cannot long continue to burn our reserves and pollute the air in search of a technological paradise.(7, 8) Others say the solution to social technological problems is more technology, not less.(9, 10) The debate will continue. Perhaps a way will be found to evolve a useful structure of values which will guide change. Perhaps we must continue to live with adversary proceedings, one position pitted against the other, always with the hope that society's good will gain in the compromise. We must leave the debate to another time.

16.4   RESPONDING TO THE CHALLENGE

The dilemma of energy demand and supply will almost certainly provide man with one of his great challenges for decades to come. We shall meet that challenge only to the degree that bright and sincere people bring their talents to bear on the problem.

For the young there is the opportunity to work within the energy and electric power industries, as engineers, planners, scientists, economists, and in many more capacities. Few industries offer such extraordinary challenges. Few industries have greater need for young ideas, for courage, for the foresight to see decades or generations into the future.

But the challenge is not for the young alone. Every citizen must participate in the great political decisions which will swirl about the growth of energy in the years ahead. The citizen will face decision after decision on atomic power, off-shore oil drilling,

FIGURE 16—2. Hydroelectric power and recreation. Shaver Lake, high in the Sierra Nevada, was formed over 60 years ago to provide dependable power to run streetcars in Los Angeles. Today it is still playing a role in power production in California. At the same time, it provides recreation for tens of thousands of people each summer. (*Courtesy Southern California Edison Co.*)

pipeline and handling facilities, automobile size and use, strip mining, energy use, and many, many more. This book has not attempted to suggest what positions should be taken. Rather it has tried to help the individual to take his stand with greater knowledge.

Some may choose conservationist positions, as espoused by organizations such as the Sierra Club, or Friends of the Earth. Others may support development of local industries, communities, and hence power companies. The choice is to the individual. Regardless of the position taken, it is essential that the citizen realize the vital role which he plays in the making of tomorrow.

16.5 BURNING BRIGHT

This last section of the book is the author's final opportunity to express his opinions.

I have tried to present as fair and complete a picture as I could of the energy crisis and the problems of electric energy generation within that crisis. I have taken the position that it is more important that the reader understand the issues than my biases. And yet it is hard to close without a last brief comment on where I believe we should go from here.

Albert Schweitzer said, "Man has lost the ability to foresee and forestall. He will end by destroying the earth." I am more optimistic. Man can be incredibly foolish, and he can be incredibly wise, when he needs to be. As the need for wisdom becomes obvious man will respond.

I believe that the balance of supply and demand should be effected by moderate cuts in demand, by reasonable energy conservation measures, and by control of new supply. We must be certain that new forms of supply are safe, reliable, and environmentally acceptable. This will take time and patience. Demand may have to be restricted to give us time. We are an energy–rich people. We can afford some dieting.

The lines by William Blake which head this chapter summarize the dilemma. In the crisis of demand and supply we have a tiger by the tail. It is a tiger which burns bright in the forest of the night: a night that was dark for so many past generations of our ancestors. This tiger was formed by no immortal hand, but by the very mortal hand of man. And it is man alone who must solve the dilemma. It is man who must decide how the resources of nature will be used.

Of all resources, the most crucial is Man's spirit.
Not dulled, nor lulled, supine, secure, replete, does Man create.
But out of stern challenge, in sharp excitement, with a burning
     joy.
Man is the hunter still,
        though his quarry be a hope, a mystery, a dream. *

*By Nancy Newhall from THIS IS THE AMERICAN EARTH by Adams &
Newhall. Copyright by the Sierra Club 1960. Used with permission.

REFERENCES FOR CHAPTER 16

1. *Electric Utilities Industry Research and Development Goals Through the Year 2000*, Report of the R&D Goals Task Force to the Electric Research Council, June, 1971.
2. L. Rocks and L.P. Runyon, *The Energy Crisis*, Crown Publishers, Inc., New York, 1972.
3. H.C. Hottel and J.B. Howard, *New Energy Technology—Some Facts and Assessments*, M.I.T. Press, Cambridge Mass., 1971.
4. R.M. Nixon, "Energy Message to the Congress of the United States," The White House, Washington, D.C., April 18, 1973.
5. H. Jackson, *et. al.*, *Senate Bill 1283*, United States Congress, Washington, D.C., March 19, 1973.
6. R.G. Stein, "A Matter of Design," *Environment*, October, 1972, pp. 17–29.
7. D. Meadows, *et. al.*, *The Limits to Growth*, Universe Books, New York, 1972.
8. B. Commoner, *The Closing Circle*, A. A. Knopf, New York, 1971.
9. P. Beckman, *Eco–Hysterics and the Technophobes*, The Golem Press, Boulder, Colorado, 1973.
10. R. Neuhaus, *In Defense of People*, The Macmillan Co., New York, 1971.

PROBLEM FOR CHAPTER 16

Write a short paper which summarizes your position concerning the dilemma posed in this book in general, and in this final chapter in particular: the dilemma of the imbalance of energy demand and supply, particularly in the area of electric energy.

INDEX